Usable Thoughts

Usable Thoughts

Climate, Water and Weather in the Twenty-first Century

Michael H. Glantz and Qian Ye

TOKYO • NEW YORK • PARIS

United Nations University Press
United Nations University, 53-70, Jingumae 5-chome,
Shibuya-ku, Tokyo 150-8925, Japan
Tel: +81-3-5467-1212 Fax: +81-3-3406-7345
E-mail: sales@unu.edu general enquiries: press@unu.edu
http://www.unu.edu

United Nations University Office at the United Nations, New York
2 United Nations Plaza, Room DC2-2062, New York, NY 10017, USA
Tel: +1-212-963-6387 Fax: +1-212-371-9454
E-mail: unuony@unu.edu

United Nations University Press is the publishing division of the United Nations University.

Cover design by Mea Rhee

Printed in Hong Kong

ISBN 978-92-808-1186-5

Library of Congress Cataloging-in-Publication Data applied for.

Dedication

We dedicate this book to the memory of William Burroughs who was responsible for preparing the book, *Climate: Into the 21st Century*, on which this publication is based.

Contents

Figures ... xi

Preface ... xvii

Acknowledgements ... xix

Abbreviations ... xxi

About this Book ... xxiii

Introduction
Rates and processes of a changing climate: The year 2020 is the new 2050 ... 1

1 **Our Perceptions of Climate** ... 7
A century of discovery ... 8
What is climate? ... 14
The challenge of measuring global climate ... 18
Climate patterns ... 22
Implications of variability ... 26
Where people live ... 30

What people think about their climate 34
The possible impact of human activities 38
Our increasing vulnerability 42
Changing sensitivity to climate 46
Climate from historical accounts 50
Twentieth century climate change 54

2 The Climate System 59
Our knowledge of climate 60
Which events matter? 64
Learning by experience; exploiting new technologies 68
Northern middle latitude winter storms 72
Drought and dust 76
Understanding tropical cyclones 80
Trends in tropical cyclone numbers 84
El Niño/Southern Oscillation (ENSO) 88
Drought in Africa 92
Human effects on climate: Urbanization 96
Human effects on climate: Greenhouse gases 100
Human effects on climate: Atmospheric particulates 104
Human effects on climate: Deforestation and desertification 108
Human effects on climate: Ozone layer depletion 112

3 Impacts of a Varying Climate 117
Suffering and costs 118
The impacts of climate fluctuations 122
Learning from experience 126
Climate and biodiversity 130
Weather, climate and warfare 134
Major temperate zone droughts 138
Storm surges 142
Tropical storms 146
Sahel – the human consequences 150
Sahel – the nature of desertification 154
Floods around the world 158
ENSO – impact on agriculture 162
Other ENSO impacts 166
Heatwaves 170
Exceptional weather and climate events 174
Freshwater: Life's precious resource 178

4 Climate for a Better Society 183
The rise of technology 184
Climate knows no boundaries 188
Serving society's needs 192

Tracking climate 196
Predicting the seasons........ 200
Food and fibre production 204
Energy and society 208
The earth's freshwater resources 212
Seeking answers on climate change........ 216

5 The Century Ahead 221
Immediate challenges 222
Adapting to future change 226
Protecting the planet and its people........ 230
Future surprises? 234

Concluding Thoughts
The Limits to Growth and *The Kyoto Protocol*: The need for "tipping point" assessments 239

Selected Reading 245

Figures

P1	Burroughs (2003), *Climate: Into the 21st Century*	xxiii
1.1	Boston (2007), *Towards a New Global Climate Treaty*...........	9
1.2	Aspects of the global climate system..................................	15
1.3	Temperature in the northern hemisphere over 1000 years....	19
1.4	Water matters!	23
1.5	Examples of human attempts to exploit the environment	27
1.6	Not everything can adapt to a changing climate. Near Churchill, Manitoba, Canada..	31
1.7	Glantz (2003), *Climate Affairs*...	35
1.8	Global warming: Causes and effects	39
1.9	Glantz (1994), *Drought follows the plow*	43
1.10	Military personnel assist hurricane Ike victims, Haiti, September 2008	47
1.11	Calligraphy in ancient Egyptian handwriting	51
1.12	Northern hemisphere temperature anomaly 1960–2005.......	55
2.1	Cellphone: Weather and climate knowledge is accessible at one's fingertips...	61
2.2	But does society know how to use it?	61

2.3 Glantz (2009), *Heads up!* ... 65
2.4 TOGA NOAA buoy ... 69
2.5 ENSO observing system ... 69
2.6 Winds in the Mediterranean region ... 73
2.7 Steinbeck (1939), *The Grapes of Wrath* book cover ... 77
2.8 1930s dust storm ... 77
2.9 A parade of storms in the Atlantic Ocean, 1995 ... 81
2.10 Media weblines for super-typhoon Maemi, South Korea, 2003 ... 85
2.11 ENSO: Degrees Celsius change from normal ... 89
2.12 Cross-section of ocean temperature in the tropical Pacific: Normal conditions ... 89
2.13 Cross-section of ocean temperature in the tropical Pacific: Developing El Niño ... 89
2.14 Cross-section of ocean temperature in the tropical Pacific: Mature El Niño ... 89
2.15 African climate ... 93
2.16 Urban heat island effect ... 97
2.17 Annual greenhouse gas emissions by sector ... 101
2.18 Gases, dust and aerosols that affect the atmosphere ... 105
2.19 Amazon fire ... 109
2.20 Ozone layer depletion causes ... 113

3.1 Refugee camp, Chad 119
3.2 Tornado illuminated by lightening........ 123
3.3 Flooding 123
3.4 Landslide........ 123
3.5 Extreme weather conditions, Latin America, 2008........ 123
3.6 Lamb (1995), *Climate, history and the modern world*........ 127
3.7 Ice cores for paleo-climate research 127
3.8 Tree rings, Historic Jamestowne, Virginia, USA 127
3.9 Earth's terrestrial (red) and marine (yellow) biodiversity hotspots 131
3.10 Retreat of the French Grand Army from Moscow 135
3.11 "The century of drought" 139
3.12 1953 Holland floods 143
3.13 Coastal dikes, Holland (improved dike protection)........ 143
3.14 Hurricane Dennis over Cuba........ 147
3.15 Famine in Ethiopia caused by drought, January 1974 151
3.16 Building barriers using local bushes and dead palm leaves to stop sand dunes from advancing on settlements, Gour, Niger........ 155
3.17 United Nations Convention to Combat Desertification (UNCCD) COP6, Havana, Cuba, 25 Aug–5 Sept 2003 155
3.18 Flooding on the Yangtze River, China 159

3.19 ENSO Impacts 163
3.20 ENSO forecast trials 167
3.21 European heatwave, 2003 171
3.22 Trends in natural disasters 175
3.23 Quantities of water required to grow common crops 179
4.1 NOAA climate services 185
4.2 Tropical storms do not respect political borders 189
4.3 Display of how climate services interact with users 193
4.4 CLIPS (Climate Information and Prediction Services) logo 193
4.5 By figuring out the workings and patterns of the climate puzzle, money can be made 197
4.6 Using climate knowledge, money can be made 197
4.7 El Niño is a relatively recently discovered phenomenon 201
4.8 Food production 205
4.9 Food production 205
4.10 Coal-burning power plants are the largest U.S. source of carbon dioxide pollution 209
4.11 Solar, wind and other renewable energy sources are future major sources of energy 209
4.12 Shiva (2002), Water wars: Privatization, pollution, and profit. The politics of water is a “hot” topic worldwide 213

4.13 IPCC (1998), *The regional impacts of climate change: An Assessment of Vulnerability* 217
5.1 The effects of climate change on rainfall in Australia 223
5.2 Work it out! COP6 227
5.3 UN General Assembly, September 2009 231
5.4 "Conference adopts Kyoto Protocol" 231
5.5 Committee on Abrupt Climate Change and National Research Council (2002), *Abrupt Climate Change: Inevitable Surprises* 235
5.6 Glantz (1999), *Creeping Environmental Problems and Sustainable Development in the Aral Sea Basin* 235

Preface

Climate change is perhaps the most important and profound adverse environmental change of our time. It is not just another change; it is the proverbial "mother of all environmental changes".

Climate change takes on great importance because of the consensus of scientists worldwide that human activities are without a doubt amplifying the naturally occurring global "greenhouse effect"; that is, carbon dioxide emissions resulting from fossil fuel burning and tropical deforestation are in large measure contributing to a warming up of the earth's atmosphere. The consequences of that human-induced global warming are potentially grave for all of humankind, regardless of who emitted what greenhouse gases and when they were emitted. While policymakers from all nations meet to discuss how to reduce their greenhouse gas emissions in an equitable way, the planet's atmosphere continues to warm. Because the various greenhouse gases have decades of lifetime in the atmosphere, however, we are already committed to a couple degrees of warming. Governments want to limit the total level of carbon dioxide so as to prevent "dangerous anthropogenic interference with the climate system", but doing so will require sacrifices that many governments are reluctant to make.

Throughout at least this century, people everywhere will have to cope with a warmer climate. This is not speculation; it is fact. Importantly, therefore, these people (alas, everyone on the planet) have to enhance their understanding of climate-society-environment interactions in order to better prepare for those changes that are known as well as those that are yet to be identified.

This book presents a collection of thought-provoking statements about various aspects of climate, water, weather and society that are taken from the 2003 World Meteorological Organization (WMO) publication *Climate: Into the 21st Century*. Following each statement (and an associated image) are brief comments based on the climate-related knowledge of the authors and their decades of experience studying climate-related impacts. They are intended to provide additional thoughts and considerations about climate-society-environment interactions so that readers can become better informed about the importance and seriousness of the potential consequences of a changing global climate system. Guided by these statements, readers are invited to ponder and to discuss with family, friends and colleagues their views on how best to ensure the future habitability of the planet and the stability of its global climate regime.

Acknowledgements

- We applaud the dedicated coordination efforts of Leslie Malone (WMO) and Mary Voice (Australia Bureau of Meteorology; retired) in production of *Climate: Into the 21st Century*, a UN World Meteorological Organization (WMO, based in Geneva, Switzerland) produced excellent illustrated book on climate edited by science writer, William Burroughs. We also give special kudos to D. Jan Stewart, Tania Sizer and Erik Halvorson for their unending support and guidance during the various drafts of *Usable Thoughts: Climate, Water and Weather into the Twenty-first Century*.

- We would like to give a sincere *merci* to Dr. Pierre Bessemoulin (Meteo-France and president of the WMO Commission for Climatology) and Mr. Lino Naranjo (MeteoGalicia, Spain) for their assistance to assure the correct portrayal of climate and weather science was displayed in this book.

- Special THANKS go to Gregory Pierce, supreme copy-editor, whose skills and critiques without a doubt improved the text and layout of this publication.

- This *Usable Science* project has in a sense been a "labour of love". To that end we set out to provide readers with a user-friendly presentation on climate-related processes, events and issues by "mining" from the WMO publication *Climate: Into the 21st Century*. We hope that we have correctly captured the spirit of Burroughs' original work.

Abbreviations

CFCs	chlorofluorocarbons
CH_4	methane
CO_2	carbon dioxide
COP	Conference of Parties
COP15	UNFCCC COP United Nations Climate Change Conference, Copenhagen, Denmark, 7–18 December 2009
ENSO	El Niño/Southern Oscillation
GHGs	greenhouse gases
IPCC	Intergovernmental Panel on Climate Change
MEDEX	Mediterranean Experiment (on cyclones that produce high impact weather)
N_2O	nitrous oxide
NOAA	National Oceanic and Atmospheric Administration
TOGA-TAO	Tropical Ocean and Global Atmosphere–Tropical Atmosphere Ocean
UN	United Nations
UNCCD	United Nations Convention to Combat Desertification
UNCED	United Nations Conference on Environment and Development

UNCOD	United Nations Conference on Desertification
UNFCCC	United Nations Framework Convention on Climate Change
WMO	World Meteorological Organization

About this Book

- The italicized text (referenced as "Burroughs (2003)") on each page introducing a new topic is taken from the book *Climate: Into the 21st Century* published for the UN World Meteorological Organization (WMO) in 2003 by Cambridge University Press.

- We hope this book will help to enhance climate, water and weather knowledge among the general public and its political representatives. Global to local climate systems constantly provide societies with lessons, that is, teachable moments. Societies need to pay attention to those lessons. An anonymous observer once noted, "Life is so hard because it gives us the test first, then the lesson."

Figure P1 Burroughs (2003), *Climate: Into the 21st Century*
Source: Cambridge University Press.

Introduction

Rates and processes of a changing climate: The year 2020 is the new 2050

The phrase "the new black" was used repeatedly in the 1980s to indicate that other colours (such as grey, brown or navy blue) had temporarily displaced the colour black's position in fashion as the versatile trend around which most other fashion accessories could be coordinated. Now it is a catch phrase used to indicate the sudden popularity of a new idea at the expense of the popularity of an older established idea. What could this fashion-related concept say about climate change?

For the past decade or so – and especially as we march into the twenty-first century – the media have warned us that the atmosphere is heating up beyond what might be expected to occur naturally. The primary cause is the build-up of greenhouse gases (GHGs) in the atmosphere, which is influencing (augmenting) the naturally occurring greenhouse effect and is the result of human activities.

Over the past few decades, scientists have learned a lot about GHGs, especially about their sources, sinks and rates of change and even about

Usable Thoughts: Climate, Water and Weather in the Twenty-first Century, Glantz and Ye, United Nations University Press, 2010, ISBN 978-92-808-1186-5

their impacts and how to monitor them. Attribution is a key problem. How do we distinguish between an impact resulting from a normally occurring change in climate to one that is the result of global warming? The problem is that no single meteorological event can be attributed to one or the other with any sense of reliability.

Though scientific findings might not yet provide a perfect picture of the future, they are certain and reliable enough to generate concern, and are thereby usable. These findings serve as a warning about foreseeable changes in the global climate system and the impacts of these changes on ecosystems and societies. If we continue on a "business as usual" path – choosing not to alter our types and patterns of energy and land use – global warming will intensify.

Scientists have developed scenarios that have generally focused on climate changes and their impacts that might plausibly be expected to occur by 2050 or 2100 if the computer model results of the Earth's changing atmosphere prove correct.

For a long time we have worried about both the projected rates of change as well as the processes of change. Many are barely discernible over short time-frames. Now, reports are coming in from scientists and the media worldwide that the rates of environmental change are occurring faster than had been previously projected for a wide range of ecological, social and climate impact factors. As a vivid example, the Arctic's ice cover is disappearing rapidly. Using sophisticated computer models, scientists had projected some years ago that there would be a notable percentage of loss in sea ice cover in the Arctic by the year 2020. Based on actual measurements, however, the sea ice melting had reached those projected levels by 2007 – 13 years early!

The rapidity of this Arctic melt has generated concern about the rates of changes occurring in other ecosystems. Around the world, warming and its impacts that had been projected to appear many, many decades into the future are emerging now before our eyes – and through the even sharper eyes of satellites and microscopes. In other words, "the future is arriving much earlier than expected".

Let's look at how the fashion concept of "the new black" relates to these impacts of a changing climate. For decades, the colour black was "in" and considered the best choice to wear since it is versatile enough to match with a huge variety of other colours and fashion accessories. Then, in the 1980s, fashion gurus anointed grey as the new black. Since then, many colours have achieved the title, including green, anointed by environmentalists.

As an analogy, we are suggesting that 2020 is the new 2050. Scenarios for 2050 or for 2100 are of much less concern to the public or to most decision makers than are scenarios closer to our contemporary time, life and governance. If science is going to be relevant to most policymakers today, then its projections must also concentrate on time periods that are far closer than those that are still half a century or more away. For those who are concerned with societal responses to a "dangerous" climate change, 2020 must be seen as the "new black".

Not only does 2020 become the new 2050, but the impacts projected for 2100, for example, may now plausibly arrive as early as 2050. Clearly, the climate is changing, and apparently far faster than we had expected.

These accelerated physical and ecological changes create a major dilemma when thinking about and acting on climate impacts since they are apparently occurring far faster than the rates at which institutional bureaucracies can effectively cope. Furthermore, because over the past couple of decades we have focused on adapting to and mitigating future impacts, we seem to have abandoned the concept of prevention.

There are at least two options available to tackle these dilemmas: (1) organizations MUST rethink their structures and functions, asking if their twentieth century (or nineteenth or eighteenth century) bureaucracies are prepared to address twenty-first century climate-related problems. Some are, but many are likely not; and (2) bring prevention back into the discussion.

Any new activities that might worsen an already existing climate change-related impact should be avoided. Note that coal-fired power plants are still being constructed, emission violations still go unchallenged, deforestation is still allowed to continue for a range of profit-generating reasons, and so forth.

We must now convince policymakers at all levels of society – from local to national to global – that 2020 is the year to fear, not 2050 or 2100.

We have often written about creeping environmental changes and problems, but we have come to realize that societies seem to have surrendered, accepting incremental, adverse environmental changes as inconsequential and of little importance to their well being. This acceptance must no longer continue. There is not much time left to combat global warming, which happens to be the most threatening of all creeping environmental changes of our times.

1

Our Perceptions of Climate

"Climate affects our lives in many ways. We are comfortable with the regularity of the seasons but dangerous events are profoundly worrying."

Burroughs (2003), p. 12.

Usable Thoughts: Climate, Water and Weather in the Twenty-first Century, Glantz and Ye, United Nations University Press, 2010, ISBN 978-92-808-1186-5

A century of discovery

Climate-related science and technology, and the expansion of information and capability for planning and early warning during the 20th century prepare us for the 21st century. Now, in the 21st century, new climate challenges confront us.

Burroughs (2003), p. 12.

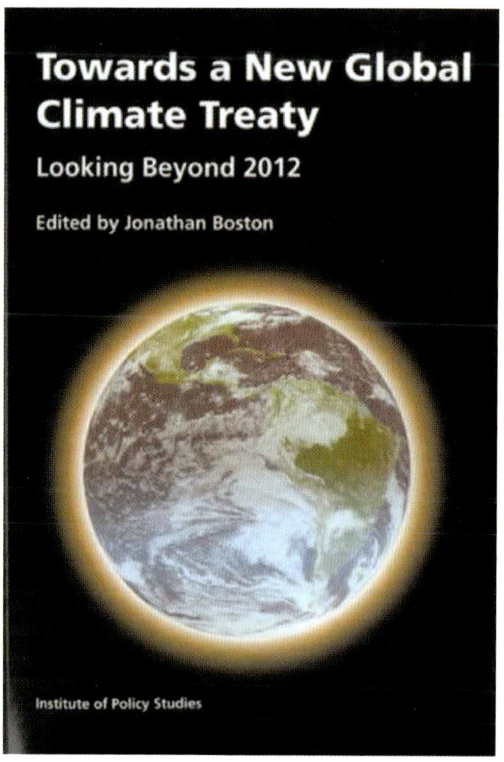

Figure 1.1 Boston (2007), *Towards a New Global Climate Treaty*
Source: Institute of Policy Studies, Victoria University of Wellington, New Zealand.

- People live and societies operate largely by the expected flow of the seasons: wet and dry, warm and cold, or various combinations of these factors.
- Most people around the globe are concerned about climate, water and weather only from season to season or when climate extremes affect them personally from time to time.
- Unbeknownst to most people is the reality that the climate system (really atmospheric behaviour) directly and indirectly influences nearly all human activities as well as all ecosystems.

- In fact, the climate system varies at all times, not just seasonally, and exhibits trends and apparent cycles in temperature and rainfall.
- Local climates that we have come to know, expect and cope with are now changing in an unprecedented way: the atmosphere is heating up with unknown consequences for lifestyles and well-being.
- Some foresee prosperity (the warmer the atmosphere, the better), while others see destitution (the worldwide collapse of ecosystems).
- The answer lies somewhere in between.

- As you read this book, please understand that no single weather or climate extreme can be blamed on global warming . . . at least not yet.
- Nonetheless, government concern about global warming of the earth's atmosphere has sharply increased, partly because unprecedented warming occurred in the twentieth century and will continue into the twenty-first century.
- Political concern centres on the extent to which societal emissions of climate-warming gases (like carbon dioxide) are a major cause of global warming.

- In 2007, several events, including the release of the Intergovernmental Panel on Climate Change's (IPCC) 4th assessment, the worldwide viewing of Al Gore's *An Inconvenient Truth*, and the awarding of the Nobel Peace Prize to both the IPCC and Gore, supplied reason enough for governments to take the causes and consequences of global warming much more seriously.
- Societies must find ways to cope effectively with a varying and changing climate and its extremes to be better prepared to cope with changes that will foreseeably accompany global warming of two or three degrees Celsius decades into the future.

What is climate?

Climate is so central to every aspect of our lives that we give little thought to what precisely it is. To appreciate fully all the reasons why the climate affects so many features of our existence, we need to define what we mean by climate.

Burroughs (2003), p. 14.

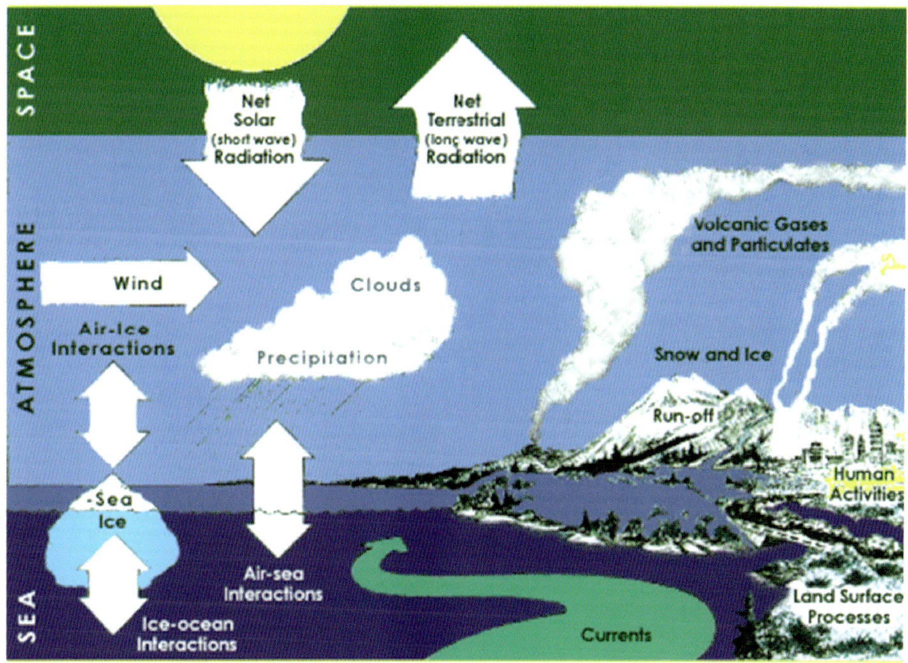

Figure 1.2 Aspects of the global climate system
Source: Government of Canada.

- Climate is really a statistical measure of weather.
  - It is simply the daily weather conditions in a given region averaged over at least a month.
  - It encompasses extremes (which are of major concern to most people), averages (means), as well as most frequent occurrences (modes) among other variables.
- Computer modellers of the global climate system use geophysical and biological factors – sea ice, forests, oceans, sun, soils, clouds, rivers, lakes, deserts, etc. – in creating their models.

- A thousand years ago these geophysical and biological factors *were* the components of the global climate system.
- Today, however, human activities are also influencing climate at local to global scales; in essence, human societies have become factors influencing atmospheric systems.
- Human activities emit gases (e.g. CO_2, CFCs, CH_4, N_2O) that influence the behaviour of the atmosphere. Societies also cut down rainforests which otherwise pull carbon out of the air. In other words, trees act as a “sink” and store carbon.

The challenge of measuring global climate

Climate data and information are increasingly used for making decisions that benefit us. Improved methodologies for agriculture, managing water resources, eradication of pests and diseases and building safer communities are some of the applications. In the past, climate data were distinctly atmospheric, but today data from the entire earth system, including the land, air, sea and ice, are needed.

Burroughs (2003), p. 16.

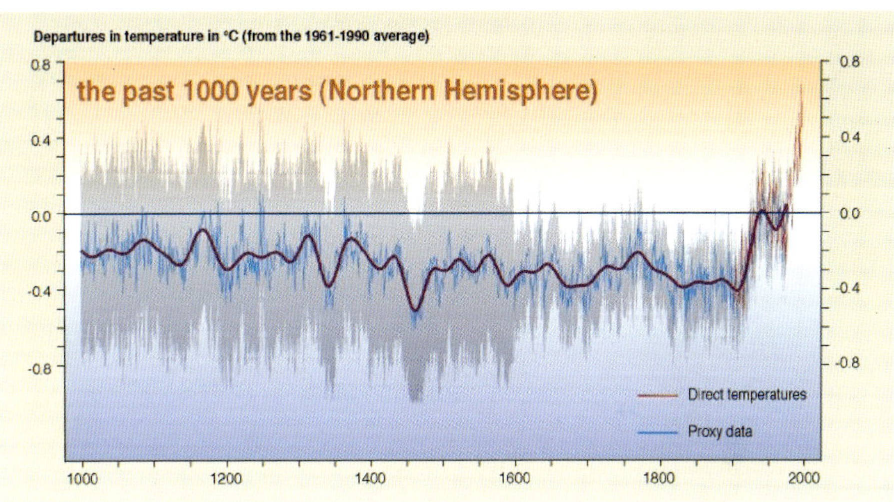

Figure 1.3 Temperature in the northern hemisphere over 1000 years
Source: IPCC (2001).

- The “hockey stick” graph on the previous page shows variations in the global temperature over the past 1000 years.
- Does global warming result from human activities, or is it a natural cyclical condition?
- Critics of global warming argue that it is a naturally occurring phenomenon and not the result of human activities such as fossil fuel burning. These people are called climate change sceptics.
- The consensus of most climate scientists, however, is that there are identifiable human fingerprints on increases in global climate temperatures.

- For many places around the globe, temperature records based on reliable instruments are relatively short.
- Therefore, to gain a glimpse of the past, researchers rely on indirect (also called proxy) information derived from tree rings, ice cores, pollen in soil sediments and so forth.
- Many historic records and much prehistoric information have been used to help researchers better understand previous climate conditions and their impacts on ecosystems and on people in settlements around the globe.

Climate patterns

Climate is not just a matter of statistics; it affects many aspects of our lives depending where on the globe we live. It is therefore equally important to define what aspects of the climate matter to us, both in human terms and to all other forms of life.

Burroughs (2003), p. 18.

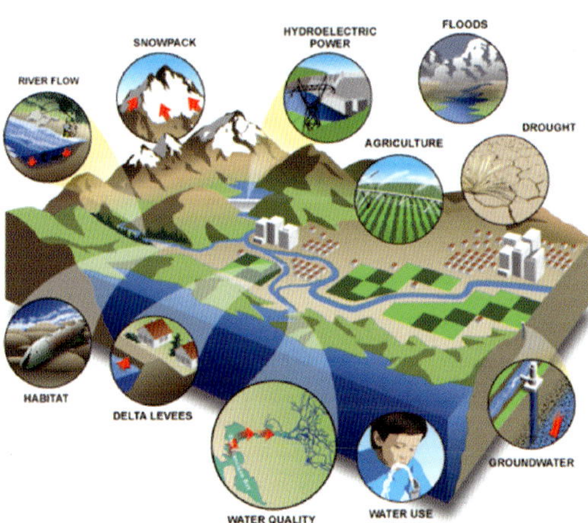

Figure 1.4 Water matters!

Source: Department of Water Resources, State of California <http://www.water.ca.gov/climatechange/factsheet.cfm>.

Climate change is already impacting California's water resources. In the future, warmer temperatures, different patterns of precipitation and runoff, and rising sea levels will profoundly affect the ability to manage water supplies and other natural resources. Adapting California's water management systems to climate change presents one of the most significant challenges of the twenty-first century.

Department of Water Resources, State of California, "Climate Change in California", June 2007, available at <http://www.water.ca.gov/climatechange/docs/062807factsheet.pdf> (accessed 29 Sept 2009).

- Defining climate is one thing; identifying the impacts of climate is another.
- Societies optimize their functioning by developing ways to cope not only with climate's variability, means, extremes and modes but also with climate trends.
- Unexpected changes in any climate characteristics will likely create both winners and losers in society.
- Few objective measures exist to indicate a climate-related gain or a climate-related loss. For example, the actual value of 10 per cent more rain in any given year in an arid zone will depend on when, where and how that rain falls.

- Climate anomalies have a ripple effect on ecosystem processes and on societal activities.
- Ripple effects can result from climate anomalies. An example is a drought that is the result of reduced rainfall having an impact on crop yields, which, in turn, lowers crop production and leads to higher prices in the market place. Ripple effects are also referred to as second- and third-order impacts.
- The general public and policymakers both need to understand how the impacts of specific climate extremes can affect the well-being of both ecosystems and societies.

Implications of variability

Climate fluctuations have an impact on human existence in many ways. The capacity of societies to adapt to extremes is finely tuned to the local climate and how much it normally fluctuates.

Burroughs (2003), p. 22.

Figure 1.5 Examples of human attempts to exploit the environment
Source: Prepared by Anne Blondel.

Each of these images represents an example of human attempts to exploit the environment in different climate zones and land surface conditions.

- Climate variations, changes and extremes affect humans from the equator to the poles and from sea level to glacier-covered mountain tops.
- People have learned over time how best to adjust to their local climates, devising ways to cope with average (normal) conditions and seasonal changes.
- When past climate extremes or changes proved too harsh to cope with, people migrated in search of more hospitable climates.
- Today, people talk about the likelihood of climate refugees. For example, citizens of the island nation of Tuvalu realize that their country could be inundated as a result of rising sea level associated with global warming. In such a case, they would be forced to abandon their country to the sea.

- Societies and individuals have always been caught off guard by surprising climate-, water- or weather-related events.
- Many societies have devised ways to withstand or rebound after these surprising anomalies.
- There have been societies unable to withstand prolonged climate anomalies, such as decades-long droughts, whose citizens were forced to abandon their lands en masse.

Where people live

We are all familiar with our local climate, so sometimes we find it hard to envisage living elsewhere, or to conceive of the consequences of our climate changing appreciably.

Burroughs (2003), p. 24.

Figure 1.6 Not everything can adapt to a changing climate. Near Churchill, Manitoba, Canada
Source: Steve Morello, Natural Habitat Adventures.

- Although people might fear the extreme future climate changes suggested by scientists, many don't think twice about choosing to move from areas with one type of climate to areas with other types.
- They either learn to live with or to accept the new climate regime and its specific extremes in the place they have chosen to live.
- This means that many people who choose to migrate from cold climates to tropical ones, or from rural areas to cities are, in fact, choosing to live under totally different climate conditions; they face different types of climate-related hazards.

- They may have had to move to a new climate because of their job or for reasons of health and welfare as a result of a climate, water or weather extremes.
- They may have been forced to change their livelihoods, temporarily or permanently, because of the negative impacts of a natural hazard like a prolonged drought, recurrent flooding, or an especially devastating cyclone.
- Being forced to migrate from one's homeland produces psychological impacts very different to those that would be experienced if the migration had been undertaken voluntarily.

What people think about their climate

How we come to terms with the everyday and the exceptional events where we live is an essential part of dealing with climate, and the better we understand our climate the better we cope.

Burroughs (2003), p. 26.

Figure 1.7 Glantz (2003), *Climate Affairs*
Source: Island Press.

Sharing knowledge worldwide about climate, water and weather science; climate's impacts on ecosystems and societies; policy and law; politics; economics; and ethics and equity is not just interesting, it is essential in order to address a wide range of social and economic problems that are climate and climate change sensitive.

- Natural science research provides people with usable information that enables them to make better informed decisions.
- But it is only partly true that the better we know the science of climate locally, the better we can cope with it.
- Social, economic and cultural factors must also be considered, and there must be a strong will to act on that information.
- Scientists now realize that they need the help of social scientists to be able to convey their scientific findings in user-friendly language to the general public.

- Most people don't choose to live in hazardous places at high risk of climate, water or weather extremes; usually, they do not have the financial means to avoid such conditions.
- Many communities in high risk areas are too poor to provide the funds needed to mitigate, adapt to or prevent adverse impacts.
- Industrialized countries responsible for saturating the atmosphere with heat-trapping carbon dioxide have a moral and ethical responsibility to assist those communities that will suffer from the social and environmental impacts of climate change.

The possible impact of human activities

The whole issue of the possible impact of human activities on the climate during the 20th century has undergone a complete turnaround in scientific attitude; from being little more than a curiosity on a local scale before the 1950s to become a dominant focus of climate studies on a global scale.

Burroughs (2003), p. 28.

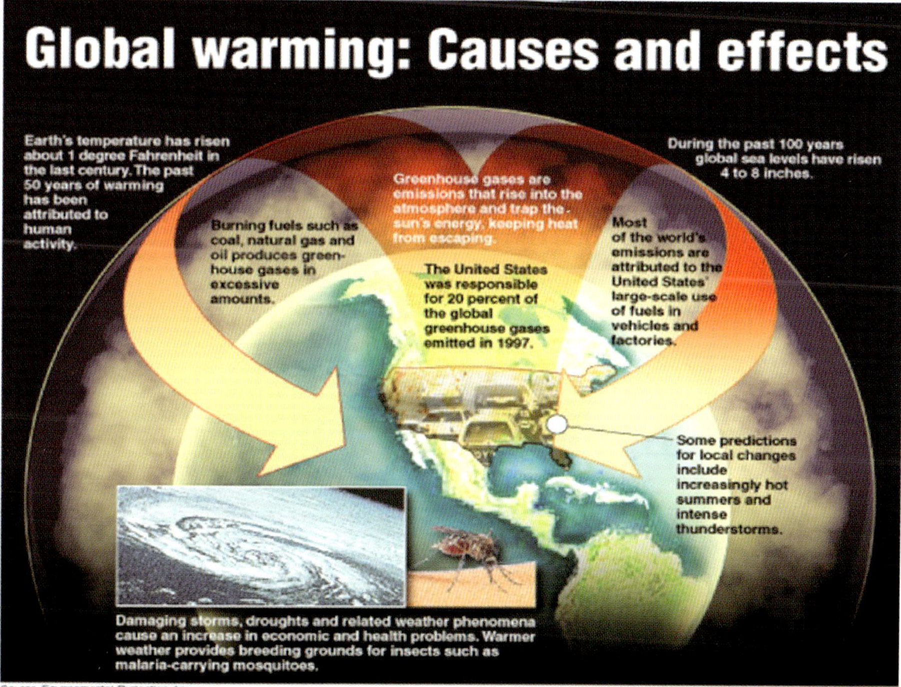

Figure 1.8 Global warming: Causes and effects
Source: Environmental Protection Agency.

- Studying climate or majoring in climatology before the 1960s was considered interesting but not very useful for day-to-day living.
- That opinion changed in the early 1970s with the occurrence of a set of anomalously damaging weather and climate events that resulted in food shortages worldwide.
- At that time, the fear was of a global cooling, not a global warming; climate scientists had identified evidence indicating a global cooling.
- Researchers at that time collected anecdotal indicators of an apparent global cooling: sea ice was appearing in ice free shipping lanes in the North Atlantic; fish populations usually caught off the northern coast of Iceland were being found along the southern coast where the water was relatively warmer; and the armadillo, which had migrated northward up to the state of Kansas in the United States, had begun to retreat southward toward Texas in response to cooler temperatures.

- By the mid-1970s, however, concern abruptly shifted toward global warming because of a build up of heat-trapping (greenhouse) gases in the atmosphere.
- Governments worldwide are now focused on global warming because of mounting scientific evidence and the awarding of the Nobel Prize to the IPCC, a global community of scientists that prepares reports on climate change, and to former US Vice-President Al Gore for his documentary, *An Inconvenient Truth*.
- Governments around the globe now take the heating of the earth's atmosphere very seriously and leaders are seeking ways to control their countries' carbon emissions and, therefore, their energy use.

Our increasing vulnerability

As the global population increases and settlements expand, many communities are becoming more vulnerable to weather extremes and fluctuations in the climate. How we minimize this increasing vulnerability will depend on making more effective use of weather and climate services.

Burroughs (2003), p. 32.

Because the best agricultural and pastural land is most likely already used, new developments have to move into land that is marginal for use under natural rainfall or soil conditions.

For example, by continuing to grow the same crops in areas becoming increasingly less productive, land degradation and crop failures are to be expected.

Figure 1.9 Glantz (1994), *Drought follows the plow*
Source: Cambridge University Press.

- Blaming the climate system for many problems that societies everywhere have to face every once in a while is too easy.
- For example, poor crop yields and production are often blamed on a lack of rainfall and water shortages, yet a closer look might indicate that certain crops are being grown in areas that are marginal, or unsuitable, for their production.
- The movement of people into marginally productive land provides an example of “drought following the plough”.

- As much as the climate itself causes problems, more and more people are moving into “harm’s way”.
- They are migrating toward coastal areas, even though coasts are known to be at risk to sea level rise and to increases in the frequency and intensity of tropical storms.
- At risk areas will likely increase in number as warming of the global atmosphere continues to go relatively unchecked.
- Climate change will surely be blamed for damage to property and loss of life when, in fact, societies and individuals must bear some of the blame. Social science researchers now devote considerable time to separating the adverse impacts that can be blamed on climate from those that can be attributed to human activities.

Changing sensitivity to climate

Whether communities are becoming more sensitive to climatic variability depends on a range of changing socio-economic factors, as well as how exposed they are to certain types of extreme events.

Burroughs (2003), p. 34.

Figure 1.10 Military personnel assist hurricane Ike victims, Haiti, September 2008
Source: UN photo/Marco Dormino.

- It is really easy to show how population shifts around the globe can increase risk, loss of life and property damage.
- It is not only a matter of population numbers but also about where people have chosen or been forced to live.
- Risks of climate, water or weather problems would likely increase even if the climate were not heating up (which it is)!
- One of the best examples of this phenomenon is the fact that people worldwide are migrating toward their countries' coastal areas, which are known to be at risk to tropical storms and storm surges even under "normal" climate conditions.

- Known hazards do not deter people from moving into harm's way.
- There are many examples of people becoming more at risk to known climate-related hazards but still choosing to move into the paths of these hazards.
- People tend to under-estimate the risks they face by, for example, moving into coastal areas or flood plains.
- People migrate from regions where they have come to know the local climate-related hazards into regions where they do not yet know the different hazards that will confront them.

Climate from historical accounts

The references to prevailing climate contained in historical writings have been difficult to interpret. How a more accurate picture of changes that have occurred has been established from these sources is an important part of how our overall understanding of the global climate system has advanced.

Burroughs (2003), p. 38.

Figure 1.11 Calligraphy in ancient Egyptian handwriting. Researchers rely on ancient records to understand the climate system.
Source: University College London <http://www.digitalegypt.ucl.ac.uk/art/writing.htm>.

To gain ideas about climate conditions in historical times when few observations were recorded, researchers rely on indirect information such as diaries, monastery records of planting and harvest times, the timing of lakes freezing, cherry blossoms and so forth.

- The lengths of most observational records for a given location are not long enough to capture the different ways that climate might vary.
- Observing current climate conditions is not enough to fully understand the global climate system.
- We must, therefore, use reliable clues (proxies) to reconstruct past climates and their impacts.
- Proxy information, and climate and weather observations taken over the past century provide the inputs to today's powerful computer models with the hope of better understanding past climates while also improving predictions of future climates.

- Therefore, we rely on historical accounts of impacts as well as on biophysical proxy indicators such as ice cores, tree rings and sediment studies.
- Together, all of this information offers us a glimpse of past climates.
- Such glimpses of the past are analogous to the paintings of the French impressionists: while the images are not perfect in their detail, they do provide a relatively good idea about our climate history.
- Over the past 1000 years, the global climate has varied naturally from century to century. Two notable anomalous periods are the centuries of warming that occurred from around the year 1000 to the mid-1300s known as the "Medieval Warming Period" and the centuries of cooling (called the "Little Ice Age") from around 1550 to the mid-1800s.

Twentieth century climate change

The global climate warmed appreciably in the 20th century. A detailed examination of the changes shows significant differences in how fast temperatures have risen during the century in different regions and where the most marked rises have occurred. These variations provide important insights into the possible causes of global warming.

Burroughs (2003), p. 40.

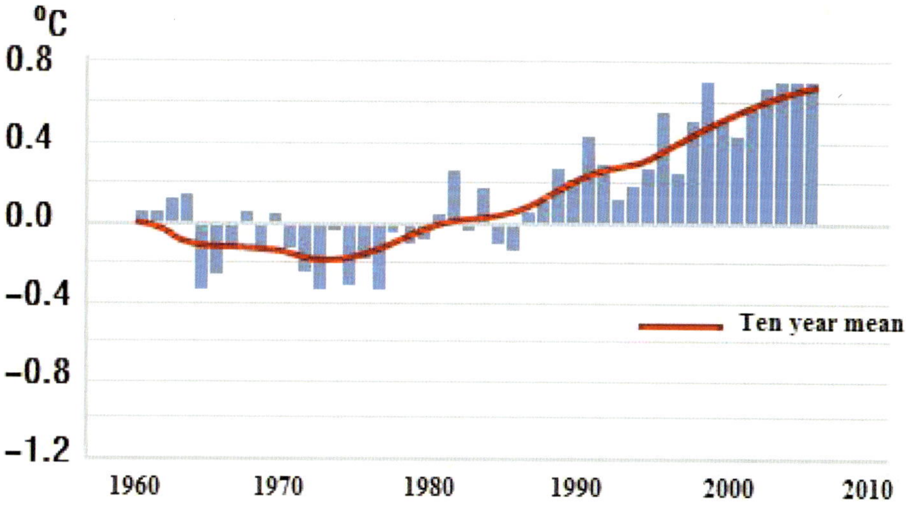

Figure 1.12 Northern hemisphere temperature anomaly 1960–2005
Source: Hadley Centre/UEA.

- In the twentieth century, the global climate heated up by 0.74°C.
- The century also witnessed runs of warm years and runs of cold years.
- The 10 hottest years on record (over 100 years) have occurred within the recent past.
- Arctic sea ice is melting at a rate faster than scientists, using their climate models, had predicted. In fact, the area of sea ice present in the Arctic Ocean in 2007 equals the amount scientists had expected there to be in the year 2020; a 13-year difference.
- Researchers in various fields are reporting that negative changes in the environment are accelerating; they attribute those changes directly to global warming.

- The debate about global climate has shifted from whether warming is happening at all to whether human activities are responsible for the warming that is happening.
- Climate sceptics challenge whether global warming is real and, if it is, whether it is a bad phenomenon. They do not accept human responsibility for warming but tend to view it as a natural occurrence.
- Evidence has rapidly increased to support the view that there is a clear human fingerprint on climate change data; however, there will always be uncertainty. But, as in life, important decisions must often be made with imperfect information.

2

The Climate System

"All aspects of the earth's climate – the wind, rain, clouds and temperature – are the result of energy transfers and transformations within the atmosphere, at the earth's surface and in the oceans."

Burroughs (2003), p. 44.

Usable Thoughts: Climate, Water and Weather in the Twenty-first Century, Glantz and Ye, United Nations University Press, 2010, ISBN 978-92-808-1186-5

Our knowledge of climate

Weather and climate knowledge is used in a wide variety of everyday services. From daily forecasts and reports, to statistical advice on one in fifty-year extremes, which need to be considered when building bridges or dams, there is abundant evidence of how much we know about the weather and our climate and how they are changing.

Burroughs (2003), p. 50.

Figure 2.1 Cellphone: Weather and climate knowledge is accessible at one's fingertips
Source: the author (Ye).

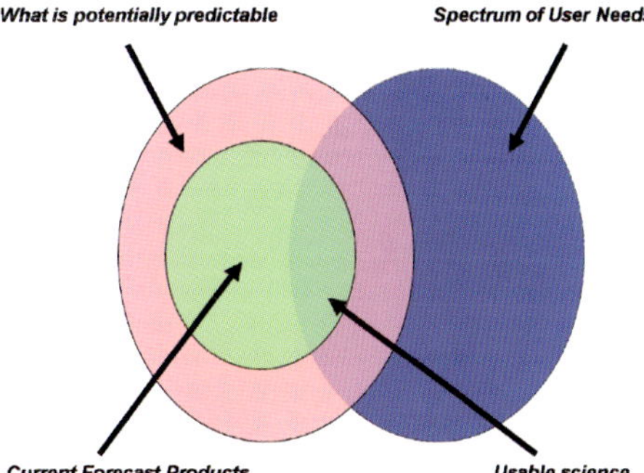

Figure 2.2 But does society know how to use it?
Source: NOAA-CIRES (2001), *Climate Diagnostics Center: 2001 Science Program Review.*

- A key word is knowledge, climate knowledge – not just data.
- Experience, qualitative information and historical accounts combine with data to produce climate knowledge for an individual, an institution and a society.
- Weather and climate knowledge can help people make better informed climate-sensitive decisions daily, seasonally and yearly.
- Most people are unaware of the extent to which climate variation and change affects their lives and well-being in both direct and indirect ways.
- Societies (both governments and individuals) must remain vigilant not only for quick-onset climate, water and weather hazards, but also for much more subtle, creeping-onset (slow and incremental, but cumulative) environmental changes that are also hazardous to societal well-being.

- Climate knowledge is also necessary for making longer term decisions that have decades-long implications, such as whether to build a factory in an area that is or may become flood-prone or to settle in an area with chronic water shortages.
- Over generations we've built up a storehouse of formal and anecdotal knowledge about weather, climate and water that we use in decision-making.
- Now that the climate is changing at a rate faster than our bureaucracies seem to be able to adjust, people will find past climate knowledge less predictive of future climate conditions that they will likely have to face. In fact, as the climate changes (however slow that may be), societies and individuals will eventually realize (with hindsight) that they had been making incremental adjustments to cope with those changes.

Which events matter?

The threats to human security from weather and climate variability come in many forms. Some, such as tornadoes, are swift and deadly while others, such as drought, are slow and insidious but in the long run can be far more damaging. The development of meteorology has been driven to a large extent by a desire to learn from the past and to minimize future human and economic disasters.

Burroughs (2003), p. 54.

Figure 2.3 Glantz (2009), *Heads up!*
Source: United Nations University Press.

Maintaining effective early warning systems is more important than many governments might realize.

- Threats to human activities and to societies can occur at different rates: slow (creeping) onset or quick onset.
- Societies have to be prepared to respond to both rates of onset.
- Changing threats require vigilance on the part of society in the form of early warning systems.
- It is sad to say that many governments do not realize the importance of early warning systems, with an emphasis on the word "systems".
- A system encompasses responses to a warning, not just to the scientific mechanism through which that warning was made.

- Information about societal responses to previous natural hazards is interesting, instructive and can provide "teachable moments".
- As the climate changes, however, the past record of impacts becomes less relevant and, therefore, less instructive.
- As a result, forecasting future behaviour of the atmosphere and its societal impacts becomes considerably difficult.
- Nevertheless, we can learn from how we responded to past climate variations and extremes in order to improve our preparations for unknown climate-, water- and weather-related hazards that might accompany future climate changes.

Learning by experience; exploiting new technologies

The development of better meteorological services has depended on improved science, the availability of new technologies, and the stimulus of events whose connection with weather and climate was sometimes less than direct.

Burroughs (2003), p. 56.

Figure 2.4 TOGA NOAA buoy
Source: NOAA.

ENSO Observing System

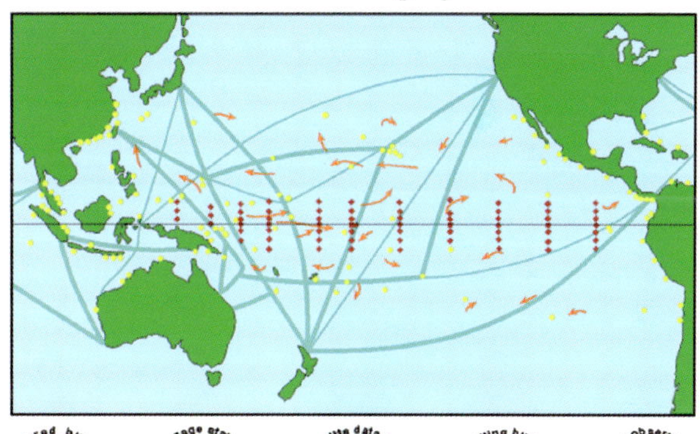

Figure 2.5 ENSO observing system
Source: NOAA, <http://www.magazine.noaa.gov/stories/images/ensoobservingsystem2.gif>.

- Many scientific advances in past decades have enabled researchers and scientists to monitor the climate system and the factors that influence it.
- Some of the best tools for monitoring the climate system have been satellites, ocean buoys and computer models.
- Yet, with all of these new and advanced technologies available, considerable difficulties remain in forecasting how the climate system will behave months to years in advance.
- Technologies are constantly being developed for data collection and analysis, remote monitoring of environmental and demographic changes, and effective communication of scientific findings to the public and decision-makers. Continual awareness of new technological developments that might be of value to improving societal understanding and use of climate-related information is imperative.

- Another tool is the Tropical Atmosphere-Ocean array developed for the Tropical Ocean-Global Atmosphere experiment (TOGA-TAO). The TOGA-TAO array of monitoring devices spans the tropical Pacific along the equator.
- The TAO array enables scientists to monitor changes in air-sea interactions as well as changes below the sea's surface that lead to El Niño and La Niña episodes.
- This information is useful not only to countries around the Pacific Rim, but also to countries around the globe whose climate is known to be influenced by changes in air-sea interactions in the equatorial Pacific.

Northern middle latitude winter storms

Throughout the middle latitudes, winter storms in the form of low pressure systems have the potential to cause widespread property damage and death. At sea, wind and wind-generated waves are the principal cause of damage. On land, wind damage and flooding are the real scourges, whereas close to shore storm surges are a major threat.

Burroughs (2003), p. 58.

There are more than 45 identified localized winds in the Mediterranean region. MEDEX provides assistance in the forecast of gale-force conditions for seven of these: Bora (Adriatic Sea), Bora (Aegean Sea), Etesian, Levante, Mistral, Sirocco and Westerly.

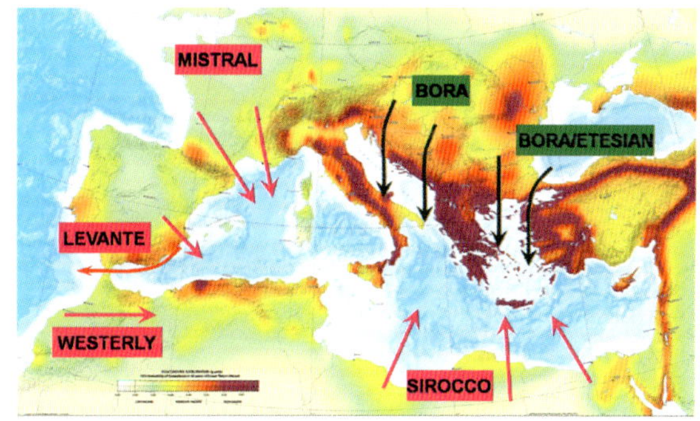

Figure 2.6 Winds in the Mediterranean region
Source: MEDEX (modified by the authors) <http://www.nrlmry.navy.mil/~medex/tutorial/medex/winds/wind_all.html>.

- Winter storms have been devastating and costly in Europe and the US.
- In 1953, the Netherlands was flooded when wind-driven waves in the North Sea overtopped the country's levees.
- In 1993, a late-winter "superstorm" plagued the eastern half of the USA.
- In 1999, thousands of trees in the park of the Chateau de Versailles in France were toppled by the worst winter wind storm in 200 years.
- In 2008, the Chinese government reported that severe winter weather – in the form of the worst winter snowstorm in 50-years – had destroyed one-tenth of the country's forest resources.

- With climate change and associated sea level rise, coastal areas will be at greater risk to storm surges.
- Sea level rise from storm surges of only tens of centimetres can cause flooding several kilometres inland.
- Extreme weather and climate events capture the attention of civil society. Societies expect their governments to protect them using science and technologically advanced warning systems.
- Nevertheless, people often have to bear some responsibility for the disastrous impacts of natural hazards because they have a responsibility to pay heed to early warnings they might receive, even if the hazards for which those warnings are issued sometimes fail to occur. Forecasts are statements of possibility and foreseeability; they do not come with a guarantee.

Drought and dust

The expansion of agriculture into the semi-arid grasslands of the world during the late 19th and early 20th centuries exposed the weaknesses of transferring agricultural practices without modification to different climate regimes. The crippling droughts of the 1930s Dust Bowl era in North America are perhaps the best known example of this bitter experience.

Burroughs (2003), p. 62.

USA, Dust Bowl Days, 1930s

Dust storms occur in many countries

Figure 2.7 Steinbeck (1939), *The Grapes of Wrath* book cover
Source: Syracuse Library Special Collections Research Center Exhibition.

Figure 2.8 1930s dust storm
Source: NOAA, <http://www.photolib.noaa.gov/bigs/theb1365.jpg>.

- When rains are plentiful, people can do many abusive things to the land that remain hidden.
- When dry spells and droughts occur, however, poor land use practices become exposed and land degradation accelerates.
- Cultivation of crops unsuitable for specific environments persists in many parts of the globe as populations move onto lands that are marginal for the production of those types of crops.
- Governments and individuals must keep in mind that over the long-term, rainfall in semiarid and arid systems is skewed to dryness and drought conditions.

- This is a fact that also must be kept in mind by people in industrialized and developing countries. A decade or so of favourable rains in an arid or semiarid system can lead governments to believe that they can exploit the soils in those systems – by converting grasslands into cultivated lands, for example – more intensively than those systems can sustain once arid conditions return.
- Recurring droughts in the 1930s – the "Dust Bowl" days on the US Great Plains – exposed poor land-use practices. Hot dry winds then carried dust all the way to the US Congress in Washington, D.C. more than 1000 miles away. As a result, the US Congress developed new land use regulations.
- The Canadian Prairie provinces of Alberta, Saskatchewan and Manitoba at the northern extreme of North America's Great Plains also suffered greatly during the 1930s Dust Bowl era.

Understanding tropical cyclones

The tropics spawn the most formidable storms on the planet. These monsters often sweep outward to higher latitudes and are an important part of the general circulation of the atmosphere in several regions in late summer and early autumn of each hemisphere.

Burroughs (2003), p. 64.

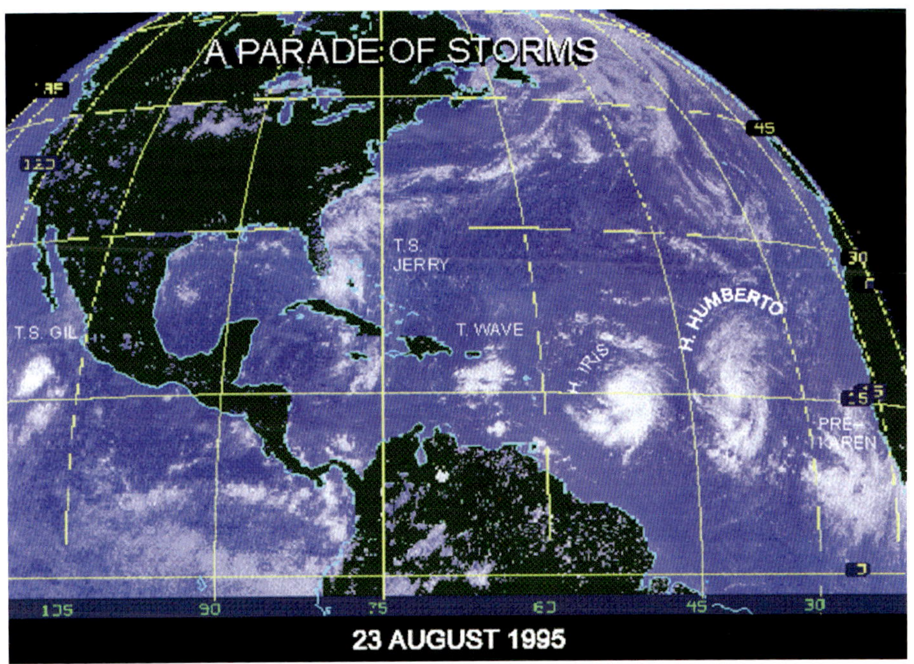

Figure 2.9 A parade of storms in the Atlantic Ocean, 1995
Source: NOAA <http://www.nhc.noaa.gov/gifs/1995mdpara.gif>.

- The satellite image on the previous page shows tropical storms lined up likely to hit the Caribbean islands and states along the eastern and Gulf coasts of the US.
- Tropical storms move heat around the atmosphere; many places throughout the region depend on them to replenish their water resources.
- The satellite image raises the fear that storm seasons in the future will be filled with blockbuster tropical storms. These blockbuster tropical storms have occurred recently with 2005 being a record-setting hurricane season in the tropical Atlantic Ocean.
- Another record-setting tropical storm year in the Atlantic was 2004, which also set records in the Pacific. Four hurricanes made landfall in the US state of Florida (the previous seasonal record had been three) and 10 typhoons affected Japan (the previous maximum had been eight).

- In the 1940s, a researcher proposed that tropical storms in the Atlantic Ocean developed off the west coast of Africa.
- Interestingly, an 1854 American scientific publication contained a map showing the tracks of numerous tropical Atlantic storms from as early as the 1790s developing along the African coast.
- Although that map did not represent systematically collected data, it is very informative and shows the value of historical information.
- This is yet another example of the value of reviewing historical climate, water and weather records, writings and other sources of information every so often.

Trends in tropical cyclone numbers

The search for knowledge of tropical storms has been driven by the death and destruction they cause. The scale of damage throughout the century has also provided the incentive to improve forecasts of their behaviour and hence improve our defences against their onslaught.

Burroughs (2003), p. 66.

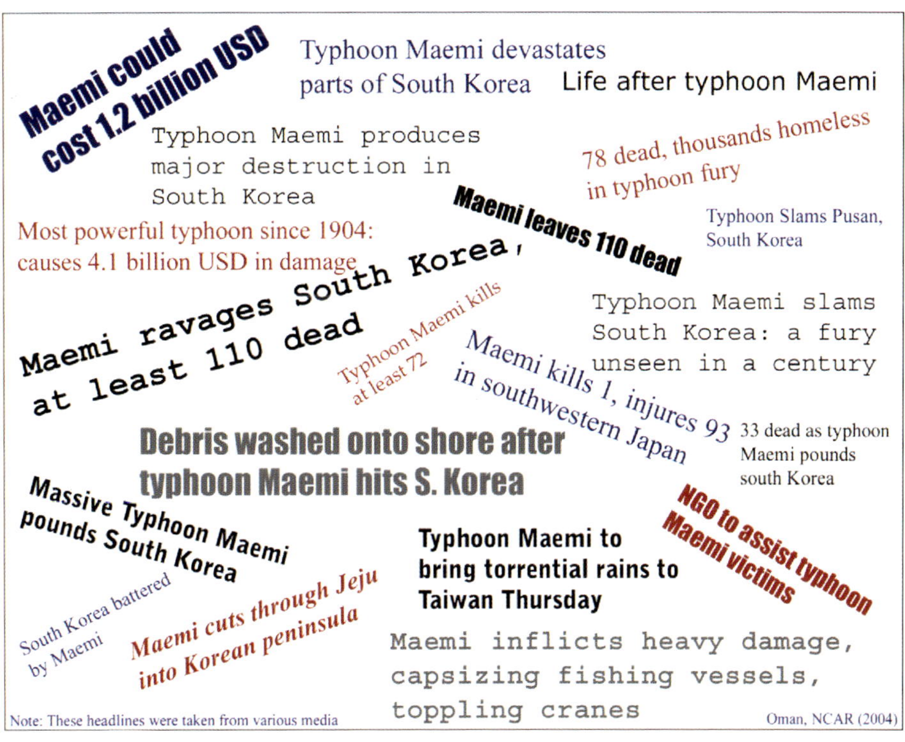

Figure 2.10 Media weblines for super-typhoon Maemi, South Korea, 2003
Source: Oman, NCAR (2004).

- Super Typhoon Maemi was a major storm that occurred in 2003 causing major impacts in South Korea. It was blamed for the deaths of more than 120 people.
- In fact, Maemi destroyed many of the cranes used to load ships in the port, curtailing harbour operations for some time.
- Interestingly, Typhoon Rusa had crossed rural Korea the previous summer and caused similar damage, but was not labelled “super”.
- Studies later showed that Maemi’s impact on the South Korean port city of Busan had been minimized because the government was able to utilize cranes from a new port facility that was under construction at the time.

- Tropical storms are the same around the world, but are called "hurricanes" in the Atlantic, "typhoons" in the Pacific and "cyclones" in the Indian Ocean.
- Storms can be deadly and destructive but do bring moisture to parched areas.
- Each year they make landfall in the US, Caribbean islands, Mexico, Central America, the Philippines, Korea, Japan, China, Myanmar, India and Bangladesh, among other places.
- Tropical storm warnings are provided by meteorological services worldwide. As Cyclone Nargis clearly demonstrated as recently as May 2008, however, an early warning days in advance of the onset of a hazard has little value if governments – in this case Myanmar's – do not wish to act on that warning.

El Niño/Southern Oscillation (ENSO)

The strongest natural fluctuation of climate on interannual timescales is the El Niño/ Southern Oscillation (ENSO) phenomenon. ENSO originates in the tropical Pacific but affects climate conditions over many parts of the world.

Burroughs (2003), p. 72.

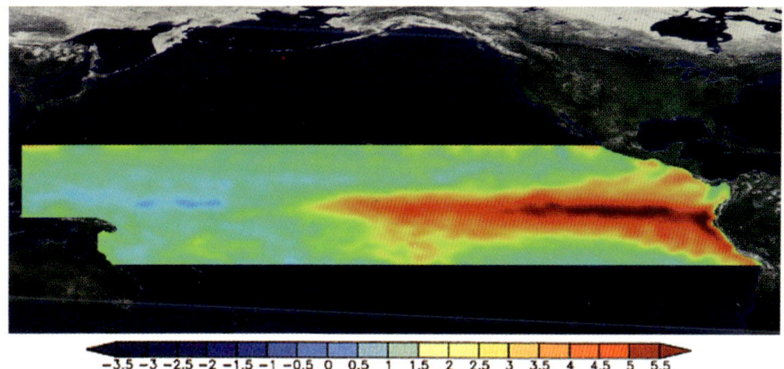

Figure 2.11 ENSO: Degrees Celsius change from normal
Source: <http://wiki.bildungsserver.de/klimawandel/upload/Enso-sst-97.gif>.

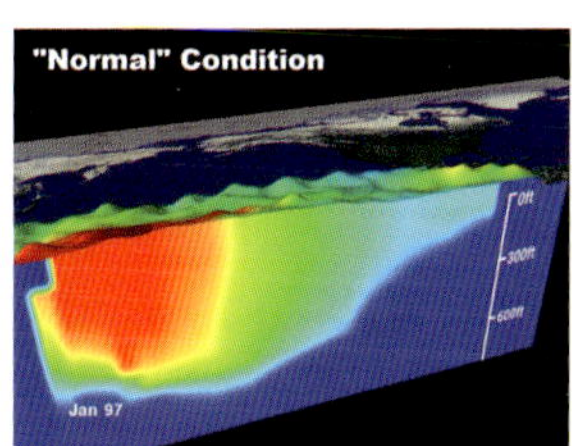

Figure 2.12 Cross-section of ocean temperature in the tropical Pacific: Normal conditions
Source: NOAA.

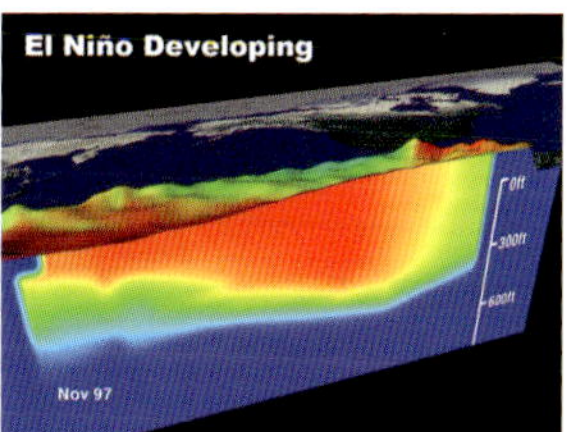

Figure 2.13 Cross-section of ocean temperature in the tropical Pacific: Developing El Niño
Source: NOAA.

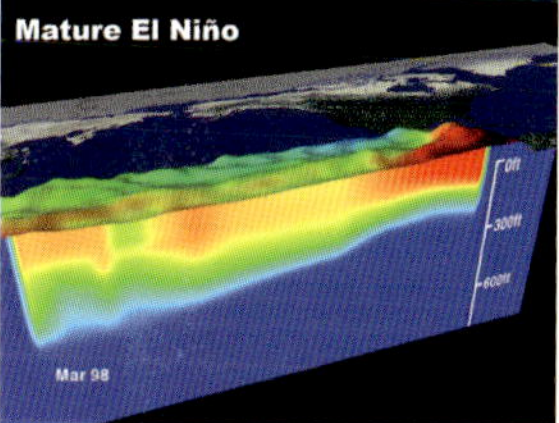

Figure 2.14 Cross-section of ocean temperature in the tropical Pacific: Mature El Niño
Source: NOAA.

- El Niño is the name given to the occasional appearance of warm surface water in the central and eastern Pacific Ocean along the equator.
- Originally, it was believed to be only a seasonal phenomenon occurring off the coasts of Peru and Ecuador with little more than regional impacts.
- Research since the last third of the twentieth century, however, has shown it to be a basin-wide phenomenon with global consequences.
- The warm water appearing seasonally along the Peruvian coast around Christmas time was named “El Niño” by local fishermen in reference to the Christ child. It would remain for a few months only to be replaced by cold, deep water welling up to the surface.

- Linkages over some distance between El Niño events and weather or climate anomalies worldwide are called "teleconnections".

- Some teleconnections are directly caused by geophysical factors in the air and sea. Others are identified by statistical correlations. There are still other impacts that come about during an El Niño event and are thought to be linked to it, though clear evidence is lacking.

- Sometimes this warm water along the coast would remain for a year or more. Eventually, the development of this more extended phenomenon was also referred to as El Niño, of which people around the globe are now aware because of its adverse global impacts. El Niño reappears on average every 4.5 years.

Drought in Africa

The causes of drought in Africa are a complex set of interactions extending throughout the tropics. The consequences are much easier to define: adequate rainfall produces healthy economies, drought brings famine and death.

Burroughs (2003), p. 80.

Warming and Water

A new study suggests that warming oceans around Africa could bring more rain to sub-Saharan Africa in coming decades, which experienced frequent droughts from the 1950s until the 1990s. But a long standing drying of southern Africa is likely to continue.

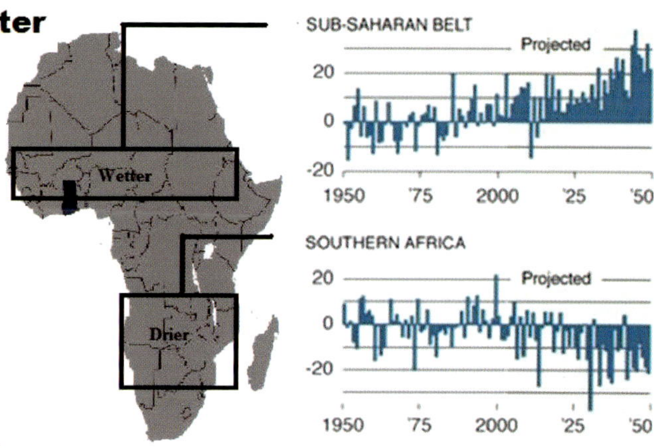

Change in mean rainfall from previous rainy season (in millmeters)

Source: Dr.Joh Eischeid, Climate Diagonstics Center, NOAA

Figure 2.15 African climate
Source: Dr. Jon Eischeid, Climate Diagnostics Center, NOAA.

- Rainfall in different parts of the African continent is variable within seasons as well as between seasons.
- Much of Africa is arid, hyper-arid and semi-arid. In such areas, rainfall is highly variable in both time and space.
- Forecasting climate from year to year is not easy, in part because of an inadequate network of rain monitoring stations.
- Sub-Saharan Africa is expected to be among the worst affected by climate change, which is particularly troublesome because African countries are already facing devastating food shortages requiring them to rely increasingly on humanitarian food assistance programs.

- Africans south of the Sahara desert suffered greatly from several multi-year droughts in the twentieth century: in the mid-1910s, 1939–42 and 1968–73.
- The 1968–73 drought in the West African Sahel was seen as part of a long-term downtrend in rainfall; drought conditions continued for several years.
- Droughts in Africa have a major impact on water supplies, seriously affecting food security on a continent that is often in need of food.
- Ethiopia, during a protracted internal war, suffered a multi-year devastating drought from 1972 to 1974. The war and the drought combined contributed to severe food shortages and famine in various parts of the country. An estimated one million Ethiopians perished during this period.

Human effects on climate: Urbanization

All urban areas alter their local climate, and the bigger the city, the bigger the impacts. Many aspects of the local weather reflect the differences between conditions in cities and the surrounding rural areas.

Burroughs (2003), p. 92.

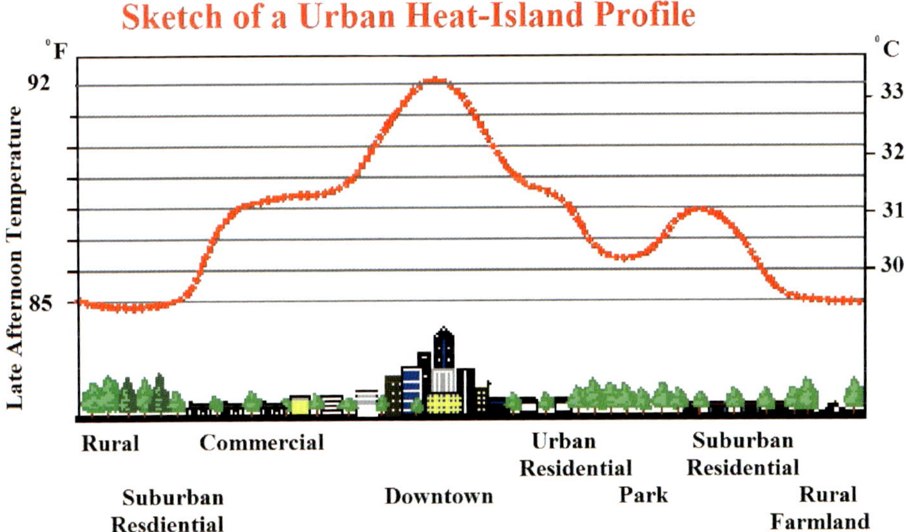

Figure 2.16 Urban heat island effect
Source: London Metropolitan University <http://www.learn.londonmet.ac.uk>.

- The urban heat island effect is an established fact.
- Cities heat up the atmosphere around them, producing special atmospheric conditions over as well as downwind of them.
- The heating of concrete and asphalt surfaces by the sun and the consumption of energy for heating and cooling add to the warming of urban areas by several degrees. This process is quite different from the global warming process.
- Urban areas are also subject to heatwaves, which are prolonged periods during which exceedingly high air temperatures put urban dwellers at high risk of heat-related stress.

- With a much warmer atmosphere, the chance of lengthy, intense urban heatwaves increases, which means that more energy is needed for cooling, thereby enhancing the urban heat island effect.
- There is a worldwide trend of people migrating to cities from rural areas in search of a better economic life.
- Are cities prepared to cope with higher populations suffering from intense heatwaves?
- Many cities have difficulty coping with heatwaves and their health impacts under today's climate conditions.
- What if global warming increases the frequency and intensity of heatwaves in overcrowded metropolitan areas where such hazards have not yet occurred? Will people in such areas be ready and prepared?

Human effects on climate: Greenhouse gases

The clearest example of an anthropogenic impact on the global climate is the change caused by the emission of radiatively active gases from human activities. Most attention has been devoted to the rise in carbon dioxide (CO_2) from the burning of fossil fuels, but a number of other emissions are also of potential importance to climate.

Burroughs (2003), p. 94.

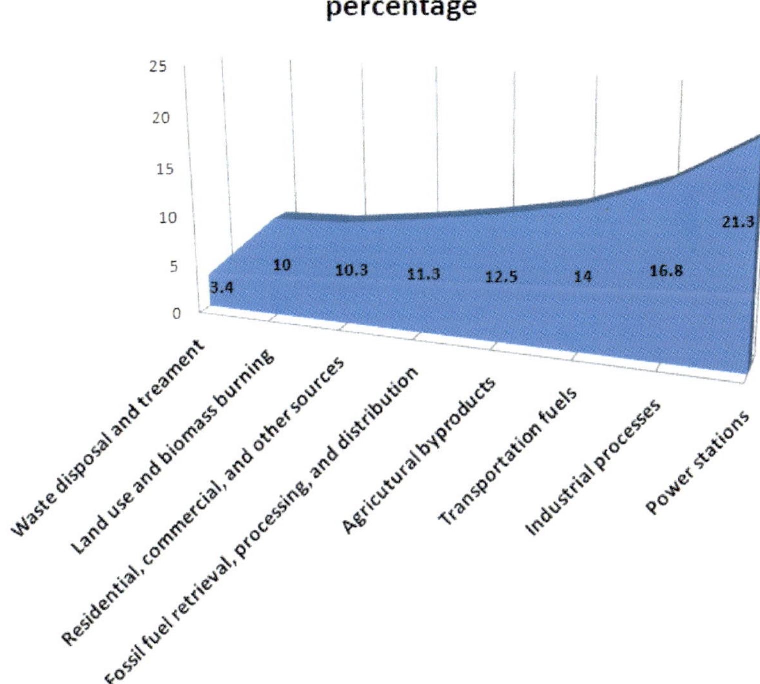

Figure 2.17 Annual greenhouse gas emissions by sector
Source: Created by the author (Qian Ye) using data from the Emission Database for Global Atmospheric Research.

- Greenhouse gases block longwave radiation from escaping the earth's atmosphere and hurling into space. They are also called heat-trapping gases.
- Global warming is often presented as a carbon dioxide (CO_2) problem; however, other heat trapping gases, such as Methane, Nitrous Oxide and CFCs, must also be considered.
- Land use also affects the processes that cause global warming. Tropical deforestation, for example, removes trees that take carbon from the air and store it for long periods of time.
- Methane, a very active greenhouse gas, is produced naturally from swamps and melting permafrost. It is also emitted into the atmosphere as a result of livestock intestinal processes; feedlots have also been identified as a major source of methane.

- Methane is 20 times more efficient than CO_2 as a radiatively active green-house gas. It retains more longwave radiation than carbon dioxide, significantly increasing atmospheric warming.
- Tropical deforestation is said to be responsible for about 20 per cent of global warming.
- Trees pull carbon out of the atmosphere as part of their growth and development. They sequester carbon; forests are carbon "sinks".
- As forests are cut down to create land for agriculture, they no longer take CO_2 out of the air. Trees that are burned return the CO_2 they have sequestered back into the air.
- Many countries have embarked on large-scale tree-planting efforts to compensate for a portion of the greenhouse gases they emit during industrial processes. This is a mitigation strategy, intending to reduce a country's total greenhouse gas emissions.

Human effects on climate: Atmospheric particulates

Sometimes termed the "human volcano", the injection of dust and liquid particles (aerosols) into the atmosphere by agricultural and industrial activities is one of the least understood aspects of the human impacts on climate.

Burroughs (2003), p. 96.

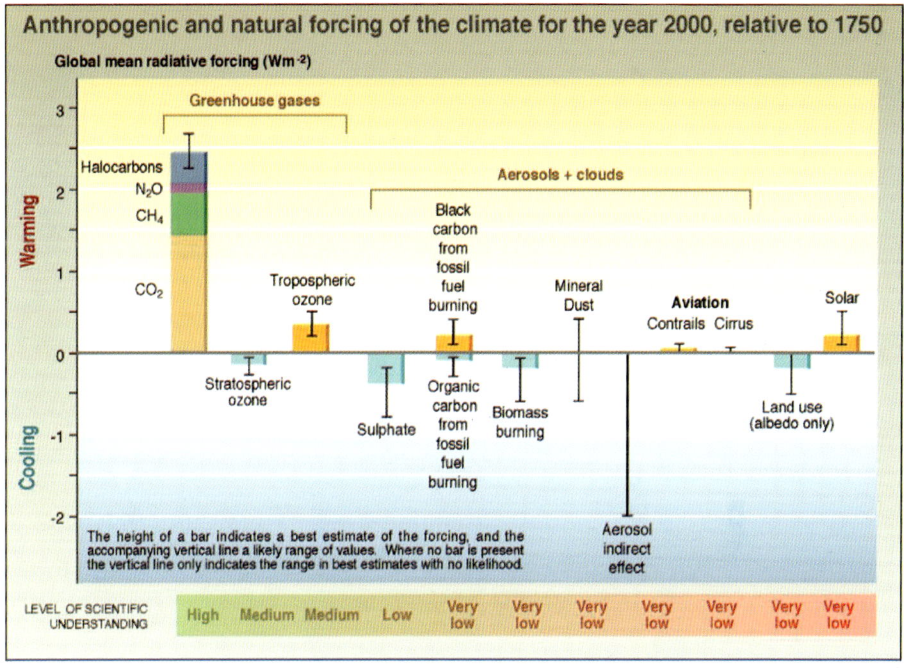

Figure 2.18 Gases, dust and aerosols that affect the atmosphere
Source: The Open University <http://openlearn.open.ac.uk>.

- Through industrialization, human activities also produce gases and particulates such as sulphate aerosols that can counter the effects of a greenhouse gas build-up; that is, aerosols have a cooling effect.
- In fact, the cooling that took place between 1940 and 1970 was in part attributed to sulphate aerosols emitted in the Northern Hemisphere during and after World War II.
- A “brown cloud” currently hangs over parts of Asia. The Asian Brown Cloud (ABC) is a two mile thick layer of pollution from Asian factories, vehicles, aerosols, forest fires and cooking fuels that blocks about 15 per cent of the sunlight reaching the earth’s surface in that region. It actually has a cooling effect on the atmosphere.

- Volcanic eruptions are known to cause a cooling of global temperatures for up to two years.
- The eruption of Mount Pinatubo in the Philippines in June 1991 is a recent example of this phenomenon.
- The eruption of Mount Tambora in Indonesia in April 1815 sharply cooled the planet, producing what was called "the year without a summer" in North America in 1816.
- Scientists are now considering ways to counter global warming by devising geoengineering schemes to mimic the cooling effect of volcanic eruptions. Eruptions inject large amounts of debris into the stratosphere and can cool the planet by more than one degree Celsius for a year or more. Scientists are proposing to use various methods to periodically inject such debris into the stratosphere. This is one of the several schemes being proposed to try to cool down the planet's atmosphere.

Human effects on climate: Deforestation and desertification

Removing and altering the vegetation cover of the land has had different impacts on the climate. These activities have been going on since the dawn of human history, but in recent decades the destruction of the tropical rain forests and changes in the extent of the world's deserts, in particular, have become major environmental issues.

Burroughs (2003), p. 98.

Amazon Basin

Rondônia

Figure 2.19 Amazon fire
Source: NASA <http://www.nasa.gov/centers/goddard/images/content/157716main_forest_burning_lg.jpg>.

About half the rain that falls in the Amazon Basin comes from evaporation and transpiration (from vegetation). Remove the rainforest and rainfall in the basin will decrease.

- Rightly or wrongly, the Amazon rainforest has been called “the lungs of the earth”, pulling CO_2 out of the air, retaining the carbon and emitting oxygen.
- The rates of tropical deforestation are very high in the Amazon Basin, Central Africa and Southeast Asia.
- The recent worldwide increase in interest in biofuels as a renewable energy resource has prompted the Indonesian government, for example, to deforest large patches of its rainforests in order to develop plantations to cultivate oil palms to produce palm oil for export to Europe and other locations.
- Other forms of deforestation, even in arid and semi-arid areas, affect local and regional climate conditions.

- Woody species are a major source of energy in the developing world. Woodcutting and overgrazing are believed to have a negative impact on regional climate conditions. Woodcutting has been called "the other energy crisis".

- The process of land degradation is called "desertification"; it can be defined as the creation of desert-like conditions where none had existed in the recent past.

- Desertification can even occur in very high rainfall areas like the Amazon Basin.

- Desertification encompasses many adverse land-use processes: livestock overgrazing, livestock trampling of soils, wind and water erosion, salinization of irrigated soils, and cutting of bushes and trees for firewood or construction of dwellings and fences (corrals) to contain livestock.

Human effects on climate: Ozone layer depletion

The destruction of the ozone layer has been a hot topic since the early 1970s. At various times scientists have pointed the finger at exhausts from high-flying aircraft, increased use of fertilizers and chlorofluorocarbons (CFCs) as being potentially damaging. We now know that CFCs have created the ozone hole over Antarctica and led to the more recent depletion in the Arctic.

Burroughs (2003), p. 100.

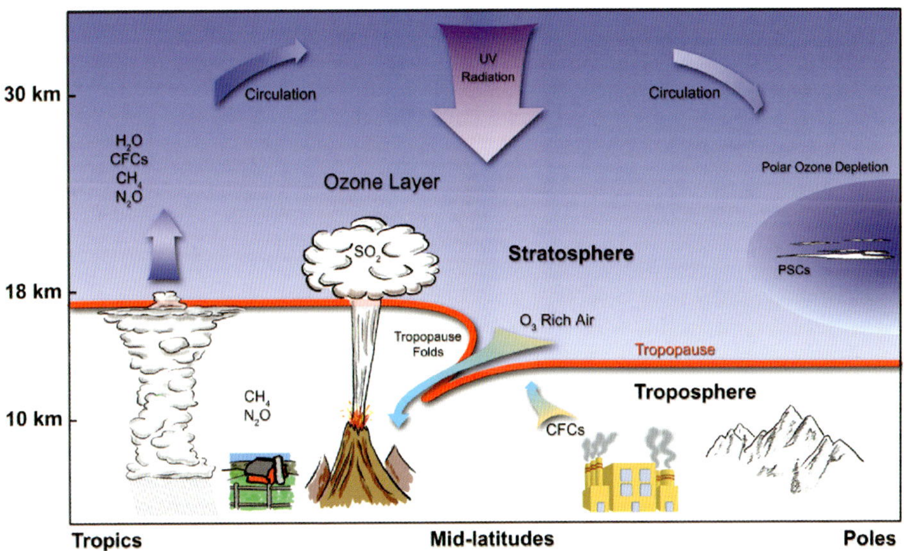

Figure 2.20 Ozone layer depletion causes

Source: NASA <http://earthobservatory.nasa.gov/Library/Aura/Aura_2.html>.

- The so-named ozone “holes” (really a thinning of the ozone layer) over Antarctica and the Arctic are the result of CFC interactions with ozone molecules. CFCs have been called “ozone eaters”.
- CFCs were once widely used as propellants in spray cans, as refrigerants, in air conditioners, as foam blowing agents and for cleaning electronics.
- The depletion of ozone in the stratosphere became a major environmental concern to scientists, the public and policymakers in the 1970s and 1980s.
- In the early 1980s, British scientist Joseph Farman identified the existence of a thinning of the stratospheric ozone layer over the Antarctic, which has come to be known as the ozone hole. This catalyzed scientists to search for the mechanisms behind the thinning and governments to act on protecting the ozone layer by reducing the use of CFCs.

- Even without direct evidence linking ozone depletion with increases in skin cancer, the "foreseeability" of that linkage was so scary that societies decided not to take chances. The production of CFCs was banned in 1987 with the signing the Montreal Protocol.
- No more than a few dozen companies were CFC producers, so it was relatively easy to ban CFC production.
- Despite the ban, however, there remains an illegal black market trade in CFCs.
- The ozone hole is going to persist for the rest of the twenty-first century, despite sharply reduced CFC emissions. CFCs have decades-long residence times in the atmosphere.

3

Impacts of a Varying Climate

"The impact of extreme weather events should not be measured only in economic terms."

Burroughs (2003), p. 106.

Usable Thoughts: Climate, Water and Weather in the Twenty-first Century, Glantz and Ye, United Nations University Press, 2010, ISBN 978-92-808-1186-5

Suffering and costs

Loss of life, permanent disruption of long-established lifestyles and setbacks to the development process carry a far greater cost to society than the matter of damage to property. Communities can be completely destroyed in climate-related disasters, an incalculable loss. Conversely, where communities recognize a hazard and band together to ward off a disaster, they can become stronger and develop more resilient ways of living.

Burroughs (2003), p. 106.

Life in refugee camps disrupts families, livelihoods and health, regardless of whether the refugees are fleeing severe drought or conflict.

Life in their villages, as they once knew it, is finished.

Figure 3.1 Refugee camp, Chad
Source: <http://www.supportdarfur.com/Darfur_refugee_camp_in_Chad.jpg>.

- When a disaster occurs, reports typically highlight the number of deaths and the cost of property destruction.
- While these are truly important factors, there is a routinely overlooked aspect of disaster – misery.
- Misery includes the loss of community (as with Hurricanes Mitch in 1998 and Katrina in 2005), the separation of families, the loss of livestock or assets, and even the loss of pets.
- While naturally occurring climate-related hazards cannot be stopped, the intensity of the ensuing disasters for society can be reduced through early warning and effective preparedness.

- Some communities have been destroyed, and others have been forced to move to avoid the impacts of future disasters.
- After each climate, water or weather disaster, lessons for doing things better are often identified. These are usually referred to as "lessons learned".
- In some places, however, disaster after disaster seem to produce the same lessons learned, suggesting that the lessons may only have been identified but not learned (or applied).
- Perhaps identified lessons should each be accompanied by a statement that outlines the likely ramifications if those "lessons learned" are not actually applied.

The impacts of climate fluctuations

Extreme weather has always had a major impact on societies. Where these extremes have bunched together over a number of years, they may be part of a longer-term climatic variation. Whatever their cause, the consequences can be catastrophic. An important question is: have the scales of economic and social impacts from climatic fluctuations increased?

Burroughs (2003), p. 108.

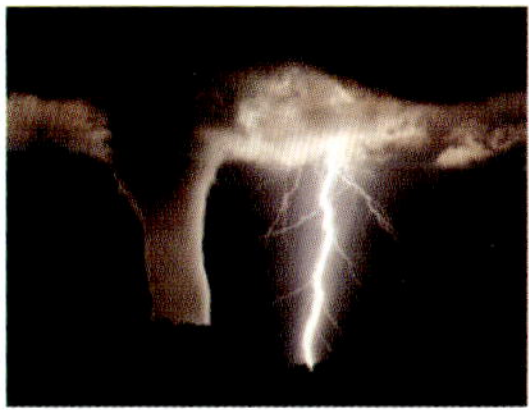

Figure 3.2 Tornado illuminated by lightening
Source: <http://www.meteorologynews.com/2008>.

Figure 3.3 Flooding
Source: Xinhua News Agency (China).

Figure 3.4 Landslide
Source: <http://www.uwsp.edu/geo/faculty/ritter/glossary/l_n/landslide.ml>.

Extremes occur every year worldwide

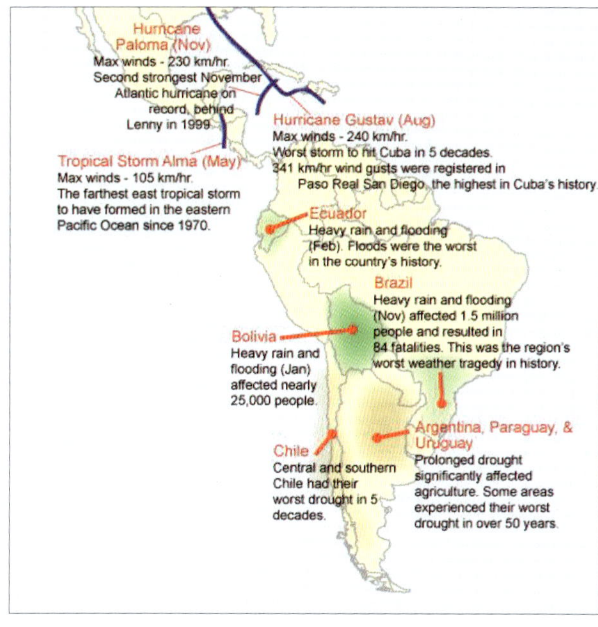

Figure 3.5 Extreme weather conditions, Latin America, 2008
Source: National Climatic Data Center, NOAA.

- Climate, water and weather extremes occur somewhere every year. Some of these extremes are likely to be memorable, record-setting events. They affect societies and individuals unequally.
- Trends in the frequency of occurrence of such extremes mean that decades-long shifts in frequency and intensity can occur.
- For example, the 1930s–60s was an active hurricane period in the Atlantic. The 1960–95 period, however, was relatively quiet. Since 1995, the number of hurricanes has again increased.
- Extreme events capture the attention of media, the general public and policymakers. The major concerns regarding climate change are whether extreme events will increase in frequency, intensity and magnitude, and whether they will occur in locations where they have not occurred before.

- Some people attribute increased hurricane activity to global warming; others say it is just natural variability. Still others recognize the influence of El Niño and La Niña in tropical storms.
- The impacts of such extremes have increased because of more costly infrastructure and because more people are moving into high risk areas.
- Today, a hurricane, typhoon or cyclone can hardly make landfall without resulting in costly impacts on land.
- It has become clear from statistical records that Atlantic hurricanes have caused fewer deaths in the past couple of decades but have been increasingly costly because of greater damage to property and infrastructure.

Learning from experience

Preoccupations with current extreme weather events make it all too easy to overlook events of the past. However, a balanced perspective of these earlier extremes is essential, both to put more recent disasters into context and to establish how effectively we have used the lessons of the past to manage more recent challenges.

Burroughs (2003), p. 110.

Figure 3.6 Lamb (1995), *Climate, history and the modern world*
Source: Routledge.

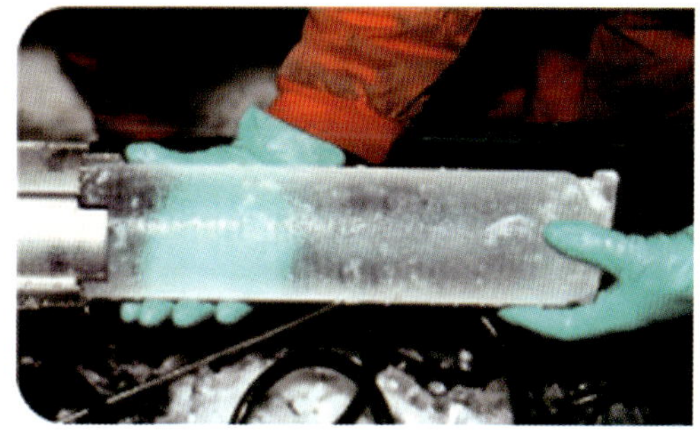

Figure 3.7 Ice core for paleo-climate research
Source: <http://oceanworld.tamu.edu/resources/oceanography-book/evidenceforwarming.htm>.

Figure 3.8 Tree rings, Historic Jamestowne, Virginia, USA
Source: <http://www.nps.gov/ner/customcf/apps/CMS_HandF/Pictures/JAME_Tree_Rings.jpg>.

- Societies must look at present-day situations in the context of the historical record; they must not only look at recent climate, water and weather events and anomalies.
- Scientists use 30-year averages (for example, 1971–2000) for research and for making projections. Each new decade prompts a recalculation of the 30-year averages of various aspects of climate (temperature, rainfall, snowpack, etc.).
- When it comes to understanding the global climate system, researchers integrate past historical information and records as well as current observations. In other words, one can see that “history has a future”! The judicious use of climate history can help to understand how societies in the past coped, for better or for worse, with climate anomalies. This in turn can help to improve coping strategies and tactics today and at least in the near to mid-term future.

- Historians and paleo-researchers have sought to reconstruct past climate-related time series for various parts of the globe.
- This data has been very instructive in identifying cooling and warming trends, and impacts in past centuries and millennia.
- Because of global warming, past climate records may become less reliable and not serve as a useful guide to decision-makers when calculating their expectations about climate impacts in the future.
- Consistent review of historical research, especially given newly developed and increasingly refined methods of analysis, could uncover insights or hypotheses of past researchers that had been overlooked or had been viewed as unimportant or untestable.

Climate and biodiversity

The diversity and distribution of all forms of life on the Earth reflect the wide range of climatic conditions in which they have evolved. This biodiversity is now threatened by human activities.

Burroughs (2003), p. 112.

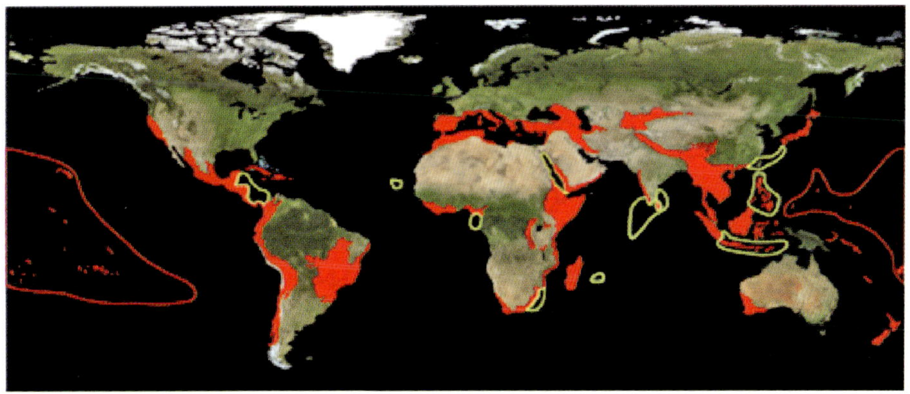

Figure 3.9 Earth's terrestrial (red) and marine (yellow) biodiversity hotspots
Source: ESA MEDIAS France/Postel.

- Many species on the earth have yet to be identified. And they may never be because natural and human-induced changes in climate are leading to their extinctions.
- Human activities involving land use also have an adverse impact on biodiversity. For example, animals need to migrate, but road infrastructure has fragmented the landscape. Climate change is hindering many species' (pests, insects, birds, mammals) ability to migrate.
- In the marine environment, heavy fishing by commercial fisheries has led to the collapse or near collapse of most species with a commercial value. In fact, highly productive coastal fisheries are under great stress due to human exploitation of technological devices that make fish populations easy to detect and capture for use as food or other products.

- With heightened concern about global warming in recent years, biodiversity issues have become a major scientific and political concern.
- For example, former U.S. President George W. Bush, a known climate change sceptic, finally at the end of his presidency recognized that polar bears are heading towards extinction as the polar regions are heating up more rapidly than other latitudes.
- The question that must be asked whenever human activities impinge on ecosystems is: who speaks on behalf of the wetlands, forests, lakes, fish populations or endangered species such as the polar bear? Who defends them against those who seek to gain benefits in the short-term without caring about the long-term ecological or societal consequences?

Weather, climate and warfare

The success or failure of many military campaigns has often rested on the ability of one side or the other to exploit weather and climatic conditions more effectively.

Burroughs (2003), p. 114.

Figure 3.10 Retreat of the French Grand Army from Moscow, intercepted by Russian Cossack, 1812, E.Orme, 1813
Source: <http://commons.wikimedia.org/wiki/File:Retreat_of_Napoleon_Army_from_Moscow_1812.jpg>.

- Some people do not believe that the weather can affect a conflict in significant ways because, so they believe, modern technology can always overcome weather constraints.
- From the beginning of settlements to the present, however, history has shown that the opposite is true.
- Settlements have been at war with a variable and changing climate since time immemorial. The challenge for societies has always been how to best survive and prosper in the face of an uncertain climate future; the most successful societies have learned how to prosper within the constraints of their regional climates.

- A classic example is Napoleon's failed invasion of Russia in 1812 at the onset of a harsh fall and winter. Hitler's attack on the Soviet Union in the early 1940s suffered a similar fate.
- President Carter's attempt to free American hostages in Iran in 1979 was foiled by seasonal dust storms; these were predictable but not taken into consideration.
- The armed conflict between Peru and Ecuador in 1997 was put on hold because of the 1997–98 El Niño. This pause in fighting led to peace talks that ended the war, an example of "disaster diplomacy".
- Regarding climate impacts on military conflict, numerous examples are notable not only in history, but in contemporary times as well.

Major temperate zone droughts

Drought has always been part of the human environment. Some parts of the world are persistently arid, but even in mid-latitudes where westerly winds provide more reliable rainfall, dry spells have, over the centuries, caused major problems for farmers and led to food shortages.

Burroughs (2003), p. 116.

Droughts past, present and future

THE INDEPENDENT

NEWSPAPER OF THE YEAR

Today .property

'I survived a mid-air crash at 37,000ft'

ONE MAN'S INCREDIBLE STORY IN EXTRA

One third of the planet will be desert by the year 2100, say climate experts in the most dire warning yet of the effects of global warming

THE CENTURY OF DROUGHT

Figure 3.11 "The century of drought"
Source: The Independent, 4 October 2006.

- Droughts capture media attention because they affect food production, supply and prices at home and abroad.
- It is difficult to tell if a dry spell will last for only a few weeks or continue and become a devastating drought.
- Dry spells have negative impacts, too. They adversely affect the activities of farmers and water supplies in the agricultural sector.
- There are different kinds of drought: meteorological drought (less rainfall than average); agricultural drought (not enough rainfall when crops need it); and hydrological drought (reduction of water flowing in a river for whatever reason).

- Just about every country has experienced one type of drought or another.
- Starvation and famine are often blamed on drought, but not every drought needs to become a famine.
- Most famines occur as a result of a combination of rainfall shortages and human factors (e.g. political or economic decisions).
- In other words, drought alone seldom produces famine. Famine is typically defined as large numbers of people facing imminent starvation. But that definition entails a rather significant temporal delay. Indeed, famine is better defined as a process that begins with reduced access to food in the marketplace because of a shortage of supply or high prices. Only when unattended does this process gradually result in widespread starvation and victims abandoning their villages en masse in search of food and shelter in refugee centres.

Storm surges

Storm surges are the curse of many coastal areas. Where shallow seas and severe storms combine, the sea can rise suddenly and sweep away whole communities.

Burroughs (2003), p. 120.

The 1953 Holland floods led to improved dike protection

Figure 3.12 1953 Holland floods
Source: Netherlands Bookseller and Publishers Association for the benefit of the Netherlands Flood Relief Fund, 1953.

Figure 3.13 Coastal dikes, Holland (improved dike protection)
Source: <http://whatiscivilengineering.csce.ca>.

- The 1953 floods in the Netherlands, pouring in from the coastal zone, were deadly and destructive.
- These floods prompted enhancements to the country's famous dike system, including dams along the coast to prevent flooding due to tidal surges.
- A large portion of the Netherlands is at or below sea level, protected from inundation only by dikes and other barriers. Its technologies and techniques to prevent flooding are being actively sought by cities like New Orleans in the US and countries like Bangladesh.
- Flooding occurred again in the Netherlands in the mid-1990s, but this time it came from the Rhine River. They had thought that the Rhine River was a managed (i.e. controlled) river.

- Today, people in the Netherlands are considering not only how to live with but also how to live on the water. They are not certain that the dike system can withstand either sea level rise or the storms that are expected to result from global warming.
- They refer to a future of living on the water as "Hydropole".
- Conceivably, millions of citizens of the Netherlands will one day be forced to relocate because of climate change induced sea level rise accompanied by an increase in the frequency of storm surges.
- Bangladesh may suffer a similar fate, which would result in tens of millions of people having to be relocated. From one of the world's most densely populated countries, where would they go?

Tropical storms

Communities around the world have developed strategies to handle the ever-present threat of tropical storms. The responses vary from place to place and reflect not only the local climate but also social conditions.

Burroughs (2003), p. 122.

Figure 3.14 Hurricane Dennis over Cuba
Source: NOAA <http://www.nnvl.noaa.gov/hurseas2005/Dennis>.

The Cuban government is effective in coping with evacuations in the face of threatening tropical storms. Though extensive damage often ensues, loss of life almost never occurs. The government uses its power to evacuate the people from at risk coastal areas.

- Tropical storms (hurricanes, typhoons, cyclones) are expected seasonally in many parts of the globe.
- The precipitation that accompanies these storms is often needed, even though too much rain sometimes falls.
- Many islands, for example, rely on tropical storms for their water supply, despite the fact that these storms often also bring destructive winds and storm surges.
- Nevertheless, tropical storms that make landfall are very destructive and tend to reverse decades of progress in economic and social development.
- There are concerns about high impact weather, water and climate events and not just extreme, record-setting tropical storms.

- High impact events also disrupt human activities and can be costly in terms of loss of life and property.
- All societies, it is safe to say, are adversely affected by high impact climate-, water- and weather-related events. For example, storms in the mid-latitudes occur occasionally, and can be deadly and destructive as well as disruptive of human activities.
- With continued heating of the global climate, scientific models project an increase in the intensity of tropical storms. An increase in the frequency of storms is also predicted. With an increase in the likelihood of two or more tropical storms wreaking havoc in a short period of time, societies will have to re-think their coping strategies. People will have little time to recover when having to simultaneously remain prepared for the possibility of another high impact climate-related event.

Sahel – the human consequences

Perhaps the most significant example of systematic and persistent regional climate fluctuation in recent decades has been a shift to drier conditions in the region of sub-Saharan Africa, known as the Sahel. Since the late 1960s this arid to semi-arid region has suffered frequent multi-year droughts which were unprecedented in the years before.

Burroughs (2003), p. 126.

Figure 3.15 Famine in Ethiopia caused by drought, January 1974
Source: UN photo/Jerry Frank.

- The West African Sahel suffered from a multi-year drought from 1968 to 1973 that led to the deaths of 200,000–400,000 people. Over 12 million head of cattle, a major asset to pastoralists and herders, were also lost.
- This particular drought highlighted inappropriate land use practices in cultivated areas and on rangelands.
- It sparked renewed concern about sands from the Sahara desert that were engulfing settled areas, causing farms and villages to be abandoned to encroaching sand dunes.
- The drought in the Sahel led to the convening of the 1977 UN Conference on Desertification (UNCOD) and to numerous studies on desertification.

- The concept of desertification as environmental degradation was first proposed in a 1949 study by French forester Henri Aubreville.
- At the time, Aubreville was writing about the destruction of dry forests in Africa, not about arid and semi-arid areas.
- Controversy still exists about whether to blame nature or poor land management around the globe as the underlying cause of desertification.
- Natural and societal processes interact in ways that make it difficult to sort out which is the dominant cause at any given time. Confounding the ability to lay blame on nature or society is the fact that not only can there be dry years in arid and semiarid areas, but there can also be dry or wet decades.

Sahel – the nature of desertification

The drought in the Sahel during the 1970s led to international action. The apparent inexorable advance of the Sahara stimulated discussion on desertification, and the extent to which it was the product of human activities, as opposed to natural climatological factors. It also resulted in international efforts by governments to curb the advance of the world's deserts.

Burroughs (2003), p. 128.

Figure 3.16 Building barriers using local bushes and dead palm leaves to stop sand dunes from advancing on settlements, Gour, Niger
Source: © FAO/Pietro Cenini.

Figure 3.17 United Nations Convention to Combat Desertification (UNCCD) COP6, Havana, Cuba, 25 Aug–5 Sept 2003
Source: Cuban postage stamp.

- Many governments, even those in tropical climates, are concerned about desertification processes.
- Although desertification occurs primarily in arid and semi-arid areas, it can occur in areas with much higher rainfall, including the Amazon rainforest.
- Desertification has been associated with sub-Saharan Africa because of the 1968-1973 drought in the West African Sahel; however, desertification worldwide is known to be an increasing occurrence.
- In the mid-1970s, the US government refused to recognize that desertification was occurring on American soil. It believed that desertification was a Third World problem, not an American one. That attitude has long since changed, but there remains a reluctance to refer to degraded land as the result of desertification.

- Desertification is both a process and an event – regardless of designation; the end point is land degradation. Such degradation can be caused by natural and human factors, such as wind or water erosion, overgrazing, wood cutting, deforestation, and soil salinization.

- Human activities can actually create desert-like landscapes in places where rainfall is high.

- While there are scores of definitions of desertification today, many observers are confused by the term. Although they might not know how to define it, they acknowledge that they would be able to recognize a desertified area if they saw one. For example, an area that has been overgrazed, deforested, salinized or a desert-like landscape.

Floods around the world

In some regions of the world, flooding is part of the way of life. Floods often cause monumental damage and loss of life. On the plus side, however, floods can be a vital source of freshwater and are an essential element of some ecosystems, replenishing nutrients in the soils, reducing the build-up of agricultural chemicals and flushing out pollutants from waterways.

Burroughs (2003), p. 130.

Figure 3.18 Flooding on the Yangtze River, China
Source: Xinhua News Agency (China).

- Flooding is generally a bad thing. It can be deadly as well as destructive of infrastructure. When floods occur, they have an adverse impact on a region's or country's economy.
- Part of the problem with flooding's adverse impacts is the result of human activities: many settlements have expanded into harm's way by moving into known flood plains or onto unstable hilly slopes.
- The devastating Venezuelan floods of 1999 took the lives of more than 50,000 people and highlighted the dangers of not respecting known constraints imposed by nature. Despite being a region with a history of heavy rainfall and unstable mountain sides, poor land use planning allowed settlements to develop in a high risk flood-plain and hundreds of thousands of people suffered the consequences.

- Under the right conditions, however, flooding can also be beneficial.

- For example, occasional flood waters usually have high sediment loads that are deposited on fields downstream, making these fields perpetually fertile because these sediments are usually rich in nutrients and minerals. The annual floods of the Nile River have made land along its river course highly productive for millennia.

- This same deposition of fertile sediments that is carried downstream in large and small rivers is known to keep farmlands fertile.

- In many locations in the tropics, people practice what is known as flood recession farming, which occurs along the banks of rivers where the stream-flow is reduced in certain seasons, enabling farmers to cultivate the seasonally exposed riverbeds.

ENSO – impact on agriculture

It was the impact of the 1982–83 El Niño event on food production around the world that first brought the global implications of the El Niño/Southern Oscillation (ENSO) to wider public notice. Once this connection was recognized, many earlier examples of this phenomenon were identified.

Burroughs (2003), p. 140.

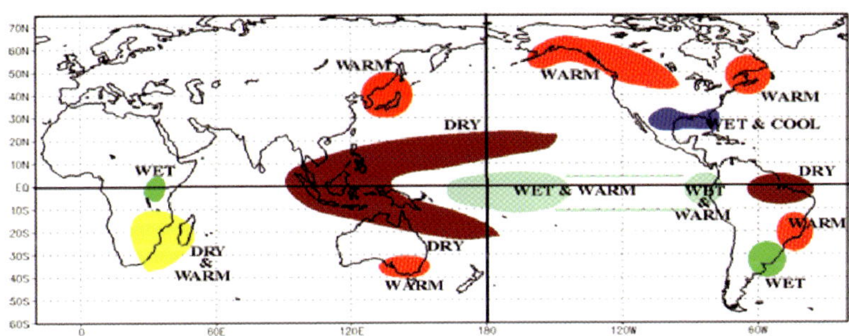

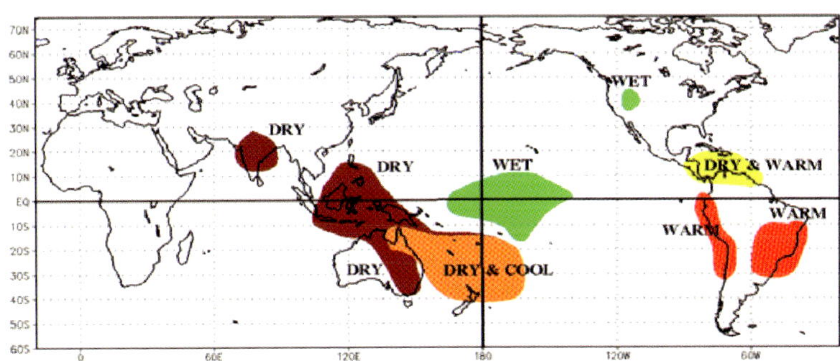

Figure 3.19 ENSO Impacts
Source: Climate Prediction Center, NOAA.

- El Niño is the result of air-sea interactions in the tropical Pacific Ocean along the equator.
- Peruvian geographers wrote about it as early as the beginning of the 1890s.
- Until the mid-1960s, El Niño was seen as a local phenomenon with local consequences.
- Until then, an El Niño event seemed to have only a slight impact on anchovy fish populations off the Peruvian coast and a major impact on guano birds that relied on these fish as a food source.
- The anchovy that is caught in Peruvian waters is used as an industrial fish. It yields fish oil and fish-meal. This fish-meal is a highly valued feed supplement used in the poultry industry. El Niño events have a major negative effect on the anchovy fish population. When this feed supplement is unavailable, the poultry industry uses other feed supplements such as soy-meal, a less preferred but less costly alternative.

- A mild El Niño in the mid-1960s provided a hint of the impact on fisheries that could be expected.
- The 1972–73 El Niño had a major impact on Peruvian fisheries, capturing the interest of several scientists. It was deemed the "**El Niño of scientists**".
- The 1982–83 event was destructive and deadly. It finally captured the attention of policymakers in many countries. This was the "**El Niño of governments**".
- The 1997–98 El Niño was the biggest in a century. Because this event made El Niño a household word worldwide, it was the "**El Niño of the people**".

Other ENSO impacts

The impacts of extreme events resulting from ENSO have, thus far, emphasized the negative aspects. By understanding the nature of the climatic anomalies in different parts of the world, however, it is possible to plan a more measured response to these extremes.

Burroughs (2003), p. 142.

The value of El Niño forecasts

Figure 3.20 ENSO forecast trials
Source: <http://www.fragilecologies.com>.

Two important things should be kept in mind about El Niños: a clever use of a reliable El Niño (or La Niña) forecast can save money as well as lives; and when an El Niño comes, some places on Earth benefit from the anomalous changes in normal weather and seasonal patterns.

- An El Niño or La Niña forecast has potential value to society, despite remaining uncertainties surrounding the science of these phenomena.
- Although El Niño is most often associated with death and destruction around the globe, it also has positive impacts.
- So, forecasts of an El Niño's onset, progress, or ending can be used by clever observers for positive purposes.
- Although El Niño returns every 4.5 years on average, it could actually reappear anywhere from two to 10 years. Because it is not an event that occurs regularly enough to give people a high level of confidence as to the timing of its onset, people tend to forget about the need to prepare for the next El Niño.
- At present, the earliest reliable warning of an El Niño's development comes only once the event has actually begun.

- For example, the normally dry but potentially fertile desert sands along the western coast of South America become grass-covered areas for livestock grazing because of El Niño-related rains.
- Fish populations also shift at these times, bringing warm water species temporarily to areas of El Niño-related warm water off the coast of Peru.
- Those who understand it are in a better position to take advantage of some of El Niño's positive impacts.
- Reducing the potential negative impacts of an El Niño event is another positive benefit of using knowledge about what an El Niño is and what it can do in various regions around the globe. Such knowledge can be used for strategic as well as tactical decision-making purposes.

Heatwaves

Heatwaves are a part of life in many areas of the world during the summer half of the year. In many cases they are no more than a nuisance, but when they reach extreme proportions they are major killers. With the threat of global warming, heatwaves may become an even greater health threat in the 21st century, especially in urban areas.

Burroughs (2003), p. 144.

Figure 3.21 European heatwave, 2003
Note: Temperature anomalies in Europe in July 2003 compared to July 2001 (the deadly 2003 summer heatwave took place in August)
Source: NASA <http://earthobservatory.nasa.gov/IOTD/view.php?id=3714>.

Summer temperatures in southern France were 10 degrees Celsius (18 degrees Fahrenheit) hotter in 2003 than in 2001. White areas show where temperatures were similar, and blue show where temperatures were cooler in 2003 than 2001.

- The meaning of global warming is exactly as it sounds: The temperature of the air around us is rising in ways that could be harmful to crops, animals and humans – especially the young, the elderly and the infirm.
- Governments want to prevent their citizens from perishing because of climate-, weather- or water-related hazards. Heatwaves cannot be stopped but their impacts can be dealt with.
- We have already seen "killer" heatwaves in the US, Europe and South Asia with hundreds to tens of thousands of deaths.
- To be sure, the young, the old and the infirm in densely populated metropolitan areas will have to take more care during high temperature episodes because such episodes are often accompanied by high levels of air pollution with its own set of adverse consequences for human health.

- The heatwave that hit Europe for weeks in the summer of 2003 was officially linked to at least 50,000 untimely deaths.
- This heatwave exposed serious weaknesses in health care warning systems and in rapid medical responses to the needs of known at risk populations.
- Today, Europe is on the alert for potential heatwaves and is seemingly better prepared to respond.
- Governments and individuals alike must remain vigilant to the risk of deadly heat. They must recognize that a human tendency is to believe that new technologies will save societies from having to suffer from impacts such as those that occurred in previous disasters.

Exceptional weather and climate events

Single extreme weather events have sometimes had a dramatic impact on societies. The full consequences can be hard to establish and often take a long time to become fully apparent.

Burroughs (2003), p. 148.

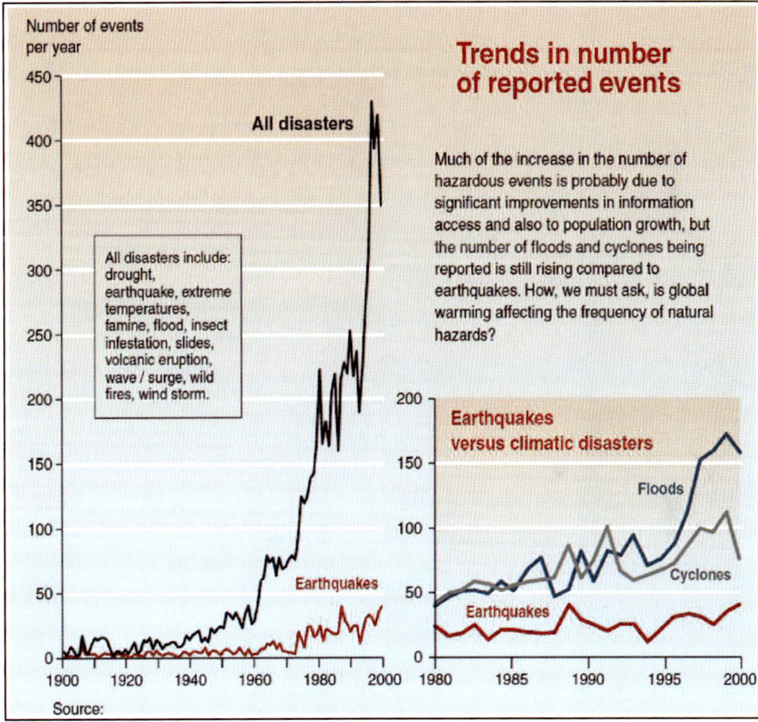

Figure 3.22 Trends in natural disasters
Source: <http://maps.grida.no/go/graphic/trends-in-natural-disasters>.

- Single, anomalous weather, climate or water events can have dramatic impacts on societies.
- People remember major droughts, floods, fires, cold waves, heatwaves, etc., especially those that caused fatalities, injuries, destruction or misery.
- These events remain memorable, even if the exact dates of occurrence are forgotten.
- In Ethiopia, for example, droughts that occurred centuries ago are still remembered and referred to by special names even though the exact years of their occurrences have been forgotten.
- In the US, people still refer to the devastation caused by the droughts of the 1930s, even though there may have been devastating droughts before and after that decade.
- Central Americans refer to Hurricane Mitch (1998) as one of its most devastating tropical storms in several centuries.

- Most people in most countries remember extreme weather, climate or water events that occurred in their lifetimes and perhaps in the lifetimes of their parents.
- Americans are sure to remember Hurricane Katrina as a devastating, even shameful, event; Indians will recall the Mumbai floods and Super Typhoon Orissa; Mozambicans will remember the typhoon that made landfall in 2000; Koreans will remember Super Typhoon Maemi; citizens of Myanmar will remember Cyclone Nargis; and so on.
- Societies must devise ways to remind coming generations about the natural hazards with which previous generations had to contend. Only then can we hope to benefit from lessons identified during previous extreme climate-related events.

Freshwater: Life's precious resource

As the world's population increases, so does the demand for freshwater. Its availability for human activity must now be carefully managed to meet basic needs and to adapt to future shortfalls caused by droughts and excess due to floods. This will require difficult policy negotiations at the local, national and international levels to reach acceptable solutions for all.

Burroughs (2003), p. 152.

Figure 3.23 Quantities of water required to grow common crops
Source: Han Jiarui (modified by the authors) <http://technology.newscientist.com/data/images/archive/2540/25401501.jpg>.

- The slogan for the UN water decade is "Water for Life". Perhaps the slogan should be "Water IS Life".
- While there has recently been a heightened focus on climate change and higher temperatures, concern is currently mounting about global warming's impacts on the future availability of water.
- A major climate concern is how the hydrological cycle will change as a result of a warmer atmosphere.
- The hydrological cycle is supposed to intensify with climate change, producing 15 per cent more precipitation, although where and when it will fall is currently unknown.

- Water, already in short supply in many parts of the globe, is becoming increasingly scarce. Some people have suggested that wars will be fought in the future over access to or supply of water.

- Concern is growing over the notion of "virtual water", which is a term used when water-scarce countries grow various products for export. Many complain when water that could be used to grow food for domestic consumption in semi-arid Kenya, for example, is used instead to grow flowers for export to the UK.

- Urban drought, as noted earlier, may prove to be the sleeper environmental crisis of the twenty-first century. With global warming, urban water supplies drawn from distant locations will become at risk to increased and more frequent shortages.

4

Climate for a Better Society

"The forces that drive the climate care nothing about territorial borders."

Burroughs (2003), p. 166.

Usable Thoughts: Climate, Water and Weather in the Twenty-first Century, Glantz and Ye, United Nations University Press, 2010, ISBN 978-92-808-1186-5

The rise of technology

At the beginning of the 20th century scientific curiosity was driving the development of technologies to better understand the climate system. By its end, awareness of the vulnerability of society was giving a new impetus to harnessing technology in all its forms to reduce risk not only from the climate but also increasingly to the climate.

Burroughs (2003), p. 156.

Figure 4.1 NOAA climate services
Source: NOAA, *Observe & Monitor* magazine.

- By the end of the twentieth century, societies had made great technological advances that enabled major improvements in forecasting various kinds of climate, water and weather anomalies and extremes.
- Interest in forecasting has been driven in part by a need to understand atmospheric processes in order to prepare for their adverse impacts as well as to take advantage of good conditions.
- People have always been intrigued by forecasting the weather and climate.
- Weather and climate knowledge also includes ordinary knowledge, such as folk wisdom that has been passed from one generation to another. A good example of folk wisdom and ordinary knowledge is the passage of knowledge about the seasonal demands of farming and rearing livestock under various types of climate and soil conditions.

- Monitoring of the atmosphere, ocean, land and sun greatly improved throughout the twentieth century.
- Scientists have picked much of the proverbial "low hanging fruit", meaning that the relatively easy findings have been discovered, leaving the more complex issues still to be understood.
- Scientists aim to uncover perfect knowledge about natural phenomena. With respect to climate, however, there will always be uncertainty. As a result, policymakers are constantly under pressure to make climate-related decisions at a given point in time with the information they have to hand. Usually this information is incomplete. But decisions have to be made. The question facing both scientists and policymakers is: When is there enough information at hand to make a reasonable decision?

Climate knows no boundaries

The forces that drive the climate care nothing about territorial borders. To deal with problems such as flooding, drought, air pollution, desertification and climate change requires information spanning countries and continents. These requirements place modern meteorological services at the forefront of international cooperation.

Burroughs (2003), p. 166.

Figure 4.2 Tropical storms do not respect political borders
Source: Japan Meteorological Agency.

- Because neither climate nor weather systems respect political borders, neighbouring countries share the impacts of their regional climate systems for the good and the bad.
- The same is also true for many rivers and river basins that cross international borders.
- International cooperation is needed to better monitor shared climate, water and weather hazards.
- International cooperation is also needed to address the potential impacts of those hazards.
- Even countries that are in conflict with one another can come together through “disaster diplomacy” in order to prepare for shared hazards. As an example, the US and Cuba, despite being known to be ideological and political adversaries, work together during hurricane seasons because they are both often impacted by the same tropical storms.

- Some countries are upstream and others are downstream.
- Downstream states are often at the mercy of those upstream because the latter can determine how and when – as well as what quality of – water will be allowed to flow to downstream users.
- Countries need to share water and weather information, not just warnings, with their neighbours.
- "Disaster diplomacy" is a concept designed to encourage neighbouring countries to take seriously the need to cooperate, despite their animosities, in order to protect or enhance the general well-being of their citizens.

Serving society's needs

As understanding of the climate system grows and society becomes more aware of the potential opportunities from the knowledge, demand develops for new and better climate services.

Burroughs (2003), p. 168.

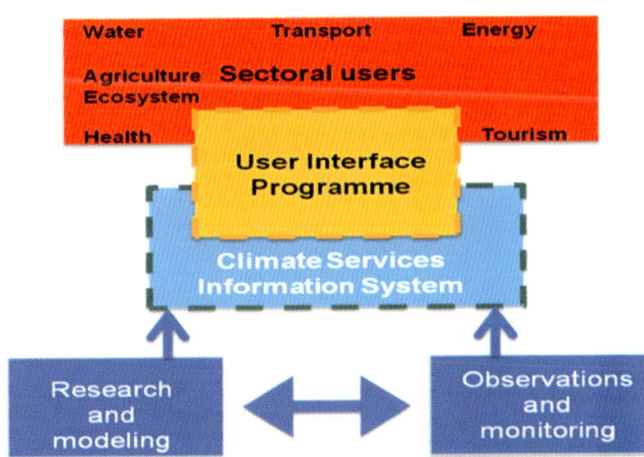

Figure 4.3 Display of how climate services interact with users
Source: WMO <http://www.wmo.int/wcc3/images>.

Figure 4.4 CLIPS (Climate Information and Prediction Services) logo
Source: WMO.

- By the first decade of the twenty-first century, societies realized that the importance of their weather bureaus rested as much on the use of their forecasts as on the making of their outputs.
- Meteorological services are broadening their activities to participate in meeting societal needs. This means they are becoming more multidisciplinary.
- Meteorological as well as hydrological services are also becoming more involved with stakeholders, that is, the users of their warnings, forecasts and reports. They now realize that there is a need to tailor their scientific outputs to meet the needs of a wide range of stakeholders' interests.

- Meteorological services in developing countries have often been operation-oriented, servicing airports and agricultural activities.
- Science acting alone, however, cannot meet the needs of most potential users of weather, climate and water information, including but not limited to forecasts.
- Considerable climate-related intellectual capacity among individuals and institutions has emerged over the last few decades in developing countries. Climate-related knowledge about agriculture, water, energy, health and public safety is central to enhancing a country's economic development prospects.

Tracking climate

Weather and climate variations have economic, social and environmental impacts at the national and regional levels and affect the course of many global markets. Managing weather- and climate-sensitive enterprises is enhanced through access to critical climate information from the past and the present, and through anticipation of future climate.

Burroughs (2003), p. 170.

Figure 4.5 By figuring out the workings and patterns of the climate puzzle, money can be made.
Source: <http://www.iahbe.org/image/money_puzzle.jpg>.

Figure 4.6 Using climate knowledge, money can be made.
Source: <http://www.money-wallpapers.com/images/money_wallpapers_23.jpg>.

- Climate, weather and water knowledge can be used to make or to save money, depending on how wise, innovative and clever the user chooses to be.
- Many examples exist of the use of such knowledge to benefit the climate-sensitive socio-economic sectors.
- Forecasts of El Niño or La Niña onsets can yield significant benefits to all socio-economic sectors of society, even those that are not directly influenced by these events.
- Collecting climate-related information about a competitor's weather and climate conditions can also provide useful information for marketing strategies and tactics.

- Commodities – such as oranges, palm oil, sugar, cocoa, grains for food and for industrial use, and fish products – are climate-, weather- or water-sensitive.
- Most individual, corporate and societal activities are geared to the natural flow of the seasons. Any disruption of that expected flow of the seasons – a longer period of summer or winter weather conditions, an early frost, or a midsummer drought – can cause major adverse impacts.
- This underscores the importance of societies maintaining effective early warning systems. As the saying goes, "to be forewarned is to be forearmed".

Predicting the seasons

Piecing together the puzzle of the ENSO phenomenon has led to the prospect of useful seasonal climate forecasts. The next step is to learn how to apply these forecasts to reduce global ENSO impacts, in such areas as fisheries failures, floods in Peru and droughts in East Africa, Australia and India.

Burroughs (2003), p. 172.

Figure 4.7 El Niño is a relatively recently discovered phenomenon
Source: J. Kitsutaka, NCAR Graphics.

El Niño is a phenomenon that was discovered as influencing global climate only as recently as the late 1960s. Many researchers around the globe are dedicated to understanding its impacts. Many aspects of its life cycle are still a mystery.

- El Niño is one of the more recent natural phenomena to have been uncovered. It has regional and global implications far from the tropical Pacific Ocean where it occurs.
- There was considerable optimism in the mid-1980s after El Niño modellers successfully made the first public forecast of an El Niño event in 1986–87.
- At first, many El Niño researchers and modellers criticized the release of this experimental forecast to the public; they later changed their views when the forecast was well-received by the media and the public, and turned out to have been correct.
- That forecast gave hope to other researchers and encouraged them to go public with their forecasts.

- Most models since the early 1990s, however, have failed to successfully predict the onset of an El Niño.
- Once an El Niño starts, it is easier to track, though scientists still do not know what makes an event begin or end or whether it will prove to be a mild or an intense event.
- Researchers are still trying to find all the pieces to the El Niño puzzle.
- The UN World Meteorological Organization now issues periodic updates on the development of an El Niño episode once the scientific community becomes concerned that one might be forming. The WMO El Niño Update is a consensus forecast prepared by scores of researchers and modellers from around the globe focused on the El Niño phenomenon.

Food and fibre production

Food and fibre production is central to human life, and agriculture and pastoralism figure prominently in the economies of most countries. Climatic data, monitoring and predictions are important ingredients for farming success.

Burroughs (2003), p. 178.

Figure 4.8 Food production
Source: <http://www.naai.moldova.md/1economy.htm>.

Figure 4.9 Food production
Source: <http://www.ilri.org>.

- Agricultural production worldwide is dependent in large measure on understanding climate variability, extremes and change.
- Climate monitoring, data collection, and analysis and prediction (including for weather and water) are invaluable for the agricultural sector – not only for one's own country, but also for countries that produce competing crops.
- Farmers in several countries are on their own when it comes to coping with a variable climate and extreme climate-related events.
- The hope is that governments will be able to provide these farmers with improved information and that information will be delivered in a timely way using a wide range of electronic mechanisms.

- Some regions have good conditions for food production. Year after year, "bread baskets" such as the USA, Argentina, Canada, the Ukraine and Australia produce surplus grains.
- This surplus is a result of favourable soils, climate and technologies that enhance productivity in the agriculture sector.
- Part of the problem with food production and access to food supplies has to do with poor distribution networks, not only in terms of transportation but also, once delivered, in terms of storage facilities.
- In many parts of the world, especially the developing world, large percentages of stored food are lost to pests and spoilage because of poorly designed storage facilities.

Energy and society

Energy is at the heart of economic and social development but is also the cause of a wide range of environmental problems, including the threat of climate change.

Burroughs (2003), p. 186.

Figure 4.10 Coal-burning power plants are the largest U.S. source of carbon dioxide pollution.
Source: <http://www.citizensclimatelobby.org/node/147>.

Figure 4.11 Solar, wind and other renewable energy sources are future major sources of energy.
Source: <http://www.in.gov/oed/images/iStock_000000584982Small.jpg>.

- Energy consumption since the onset of the Industrial Revolution in the mid-1700s has been responsible for emissions of carbon dioxide that are causing an increase in global atmospheric temperatures.
- The burning of coal has led to atmospheric pollution and associated health effects, acid rain and now global warming.
- Knowledge about the impact of higher concentrations of carbon dioxide in the atmosphere as a result of burning coal is not really new news. Swedish Nobel Prize recipient Svante Arrhenius, as early as 1896 and later in 1906 and 1908, calculated the impact of wide-scale coal burning on the global climate. At the time, a warming of the earth's atmosphere was viewed as a good thing because a warming would avert the onset of the next Ice Age. This view persisted until at least the 1940s.

- There is a lot of interest in how to reduce carbon emissions by using alternative sources of energy like solar, wind and biofuels. Nuclear energy has also reappeared as an option.
- Biofuels are controversial because farmers use agricultural commodities to produce energy for cars instead of food for consumption. This increases the likelihood of food shortages and higher food prices.
- Nuclear energy remains controversial in many countries due to a fear of potential nuclear facility accidents or near accidents, and the lack of effective nuclear waste disposal mechanisms or sites. No one seems to want nuclear waste buried in their backyard.

The earth's freshwater resources

Half of the world's people live in cities and rely on distant sources for their water supplies. There is an escalating freshwater crisis due in large part to water shortages, pollution and ineffective sewage management.

Burroughs (2003), p. 188.

Figure 4.12 Shiva (2002), *Water wars: Privatization, pollution, and profit*. The politics of water is a “hot” topic worldwide.
Source: South End Press.

- People are increasingly migrating to cities and the concentration of urban populations is constantly rising.
- As populations increase, the pressure on water resources increases, even under normal climate conditions.
- When anomalies occur, cities rely on sometimes voluntary and sometimes mandatory rationing.
- Urban droughts could become the "sleeper" crisis of the twenty-first century.
- It is possible that in years to come urban centres will have to limit the size of their populations based on their future and perhaps limited access to water resources.
- Can you envision a roadside sign of the future saying "Shanghai is full, please go to Nanjing"?

- With global warming, shortages of clean water from the tap could increase in cities.
- Access to clean water is increasingly becoming problematic; hence, the meteoric growth in the sale of bottled water.
- Coastal urban areas, especially megacities on the coast, are at risk because of storm surges. Many are also situated on estuaries whose sources begin in distant upstream areas.
- Saltwater intrusion into freshwater aquifers also presents a challenge to inhabitants of coastal cities. As climate change-induced sea level rises, the incidence and likelihood of saltwater intrusion will increase. This process is already occurring in estuaries when droughts in the interior of a country cause a reduction in the amount of water in rivers that flow into those estuaries.

Seeking answers on climate change

During the past two centuries our success as a species has created a first major global environmental problem. Now we understand many of the causes. Importantly, people want to know whether characteristic features of the climate where they live, and to which they have become adapted, are going to change, and whether there is anything they can do about it.

Burroughs (2003), p. 198.

Figure 4.13 IPCC (1998), *The regional impacts of climate change: An Assessment of Vulnerability*
Source: Cambridge University Press.

- In the mid-1950s, scientists referred to the build-up of human-induced greenhouse gas emissions as a neutral experiment that societies were performing on the earth's atmosphere. The outcome of that experiment was considered, at the time, to be uncertain.
- Researchers now know that human activities are linked to the heating of the atmosphere. In other words, it is no longer an experiment.
- Also evident is that societies are now, in the twenty-first century, an integral part of the global climate system. It is patently clear that human activities now influence the atmosphere at local to global levels. In a sense, societies are like sea ice, and forests and clouds in that they also affect the behaviour of the atmosphere.

- Policymakers, corporate leaders and the public now accept that global warming is a very threatening environmental problem.
- Industries have become so dependent on fossil fuels, that there is no simple way out of the global warming problem. There is no “silver bullet” quick fix to stop global warming.
- We must ask ourselves, “Is the climate change problem really an energy problem?”
- If the answer to this question is “yes”, then we must address the heart of the climate change crisis: the dependence of many societies around the globe, especially industrialized ones, on greenhouse gas producing fossil fuels (coal, oil and gas).

5

The Century Ahead

"The technical progress of the last century, together with growing understanding of the climate system, will be carried forward in the 21st century."

Burroughs (2003), p. 202.

Usable Thoughts: Climate, Water and Weather in the Twenty-first Century, Glantz and Ye, United Nations University Press, 2010, ISBN 978-92-808-1186-5

Immediate challenges

The most immediate climatic threats to humankind relate to increased variability in storm and rainfall patterns, more heatwaves in major urban areas, and the impact of rising sea levels on low-lying coastal regions. How we face up to these challenges may define our ability to mitigate longer-term changes.

Burroughs (2003), p. 210.

If societies cannot deal with extreme weather conditions today, how can they be expected to in a warmer climate regime? Societies must develop effective coping tactics and strategies as they head toward a future of climate-related extremes.

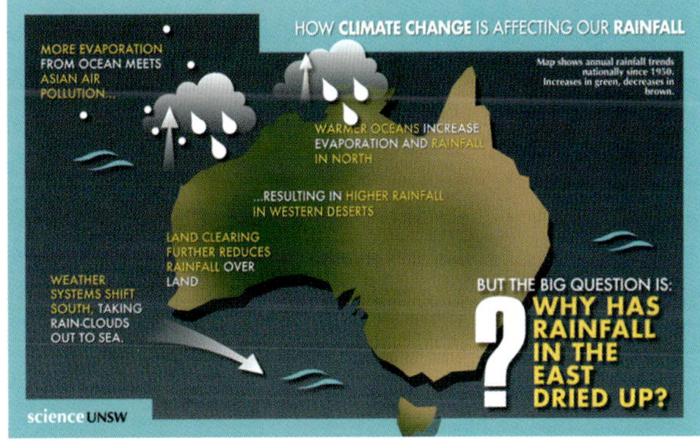

Figure 5.1 The effects of climate change on rainfall in Australia
Source: <http://www.environment.unsw.edu.au>.

- Scientists use computer models to generate scenarios about the possible implications of a warmer atmosphere. Suggested impacts include:
  - Increasingly intense and more frequent severe storms;
  - More droughts, floods and fires;
  - A poleward movement of tropical diseases;
  - Increasingly severe hurricanes, typhoons and cyclones that can be viewed as “superstorms”;
  - Changes worldwide in the expected flow of the seasons.

- Improving how we deal with extremes today can help to improve how we might cope in the future with climate change-related extremes.
- Societies learn from each present-day extreme about what is a best practice for coping with climate-related extremes. These events are "teachable moments".
- Because populations are increasing and shifting, and because the climate is changing, "best practices" and "lessons learned" from the past must be increasingly used with care. Best practices can serve us well in the short-term but they may become less applicable in the long-term.

Adapting to future change

Successful adaptation to climate change will depend on the validity of predictions and on sound political and economic policies implemented at national and international levels.

Burroughs (2003), p. 214.

Figure 5.2 Work it out! COP6
Source: <http://www.iisd.ca/climate/cop6/pix/1ipcc.jpg>.

- Adaptation to climate change and its impacts is apparently the preferred societal response to global warming.
- Adaptation as a concept has been captured and redefined to meet the concerns of those seeking policy responses to climate change.
- Some scientists refer to adaptation as a response only to those impacts on society that can clearly be shown to have resulted from climate change.
- Others refer to adaptation as any response to a variable and changing climate, whether or not linked to climate change.

- Resilient adaptation is an attempt to identify a flexible, incremental adjustment to a changing climate, the impacts of which will reveal themselves in yet to be identified ways.
- Proactive resilient adaptation involves anticipating the impacts of climate change and making adjustments to human and societal behaviour.
- Prevention must also be an explicit strategy; activities that are known to have negative impacts can be stopped in their planning stages.

Protecting the planet and its people

International cooperation is the only way to produce a coherent response to global climate issues. How nations participate in international activities, and how they design their own national programs to fit in with these efforts, will determine how successfully we face the challenges.

Burroughs (2003), p. 216.

Sometimes . . . we have to act global.

This is one of those times!

Figure 5.3 UN General Assembly, September 2009
Source: UN Photo/Eskinder Debebe.

The Japan Times

Conference adopts Kyoto Protocol

Greenhouse gas emissions to be cut by average of 5.2%

KYOTO COP3

Kyoto convention's success open to debate

Cost of Kyoto Protocol

Figure 5.4 "Conference adopts Kyoto Protocol"
Source: The Japan Times, 12 December 1997.

- Global warming requires a global response because all countries contribute greenhouse gases to the atmosphere.
- To date, industrialized countries have contributed most to carbon dioxide in the atmosphere. Now, graduated developing countries like China, India and Brazil are starting to rival developed countries in total CO_2 emissions.
- CO_2 emissions per person, however, present a different picture. China has a low emissions per capita value, whereas the US and Europe have high per capita values.

- Each baby born in an industrialized country today will, in its lifetime, use up to 60 times more energy than a baby born in a developing country.
- Those countries that have been saturating the atmosphere with excessive emissions of CO_2 since the eighteenth century have a moral responsibility:
  - to reduce their CO_2 output;
  - to provide clean energy technology to developing countries. To do otherwise would be to punish the victims of global warming's impacts, specifically in developing countries.

Future surprises?

Many aspects of future climate will surprise us. The changes, however, will only be part of a variety of unexpected developments concerning the environment and, more generally, wider economic, political and social developments. We will need to keep a sense of perspective about how the climate fits into this array of challenges.

Burroughs (2003), p. 218.

Will the global climate change with a bang or a whimper?

Figure 5.5 Committee on Abrupt Climate Change and National Research Council (2002), *Abrupt Climate Change: Inevitable Surprises*
Source: National Academies Press.

Figure 5.6 Glantz (1999), *Creeping Environmental Problems and Sustainable Development in the Aral Sea Basin*
Source: Cambridge University Press.

- Consider the rates of climate change as being abrupt or creeping. Governments do not cope well with either extreme rate of change, so they tend to call for more science when confronted with extremes, which of course allows them more time to pursue "business as usual" strategies.
- For societies, rates of change are as important as magnitudes of change.
- Researchers know a lot about the global climate system. Even with their climate knowledge, however, they do not know about thresholds of change.

- Scientific uncertainty should not be used as an excuse to continue likely harmful-to-the-climate-system human activities. Application of the “precautionary principle” is warranted.
- If there are unknown thresholds of change, it would be wise to approach the future with caution.
- Foregoing short term profit in the name of precaution to favour environmental protection should be the guiding rule for governments, corporations, groups and individuals.

Concluding Thoughts

The Limits to Growth and *The Kyoto Protocol*: The need for "tipping point" assessments

In 1972, a two-year study called *Limits to Growth*, commissioned by the influential Club of Rome, was published. College students today have probably never heard of this study, but it was one of the most influential of the time, an early attempt at using computer projections to calculate resource availability and use worldwide. The study concluded that in a hundred years or so from its release the global community would have consumed many of the planet's natural resources. It did, however, also acknowledge that its findings were preliminary and that changes in patterns of consumption could affect – either positively or negatively – the longevity of the earth's natural resources.

The study was attacked from many sides: its assumptions were challenged as being unrealistic because, for example, it noted at a time when people did not believe that the planet's resources could ever be at risk of running out that many natural resources were finite (N.B. today, experts talk about "peak oil", meaning that oil is near the end of its availability

Usable Thoughts: Climate, Water and Weather in the Twenty-first Century, Glantz and Ye, United Nations University Press, 2010, ISBN 978-92-808-1186-5

in quantities required by the global community to meet its development objectives). The study authors' trends in the rates of resource consumption by a rapidly growing global population were also challenged. Even the authors' credibility was under attack. Nevertheless, an estimated 50 million copies of *The Limits to Growth* were sold.

20 years later, the authors of *The Limits to Growth* wrote a sequel entitled *Beyond the Limits*, in which they commented,

> [*The Limits to Growth*] created a furor. The combination of the computer, MIT, and the Club of Rome pronouncing upon humanity's future had an irresistible dramatic appeal. . . . Our book was debated in parliaments and scientific societies . . . *The Limits to Growth* inspired some high praise, many thoughtful reviews and a flurry of attacks from left, right and middle of mainstream economics . . . The book was interpreted by many as a prediction of doom, but it was not a prediction at all. It was not about a preordained future. It was about a choice. (Donella Meadows et al., *Beyond the Limits*, 1992, p. xiii).

The legacy of *The Limits to Growth* is amazing. It not only stimulated pro and con discussions, but more importantly it spawned hundreds of reports worldwide about limitations to resource use and to economic growth. These include: *Mankind at the Turning Point* (1974), *Global 2000* (1980), the *Brandt Commission Report* (1980) and the *World Commission on Environment and Development* (*Our Common Future*, 1987), among others.

Mention of *The Limits to Growth* still raises praise in support of its findings or eyebrows in opposition. The truth is that the authors, at the time, had explicitly acknowledged that their study was neither perfect nor complete and that there was hope to change the ways that natural resources around the globe were being used. As noted in *Beyond the Limits*:

> In [*The Limits to Growth*] we raised questions such as: What will happen if growth in the world's population continues unchecked? What will be the environmental consequences if economic growth continues at its current pace? What can be done to ensure a human economy that provides sufficiently for all and fits within the physical limits of the earth? (p. xii)

Fast forward to the late 1990s and to the Kyoto Protocol. In our view, the Kyoto Protocol was crafted in 1997 as an expression of growing international support to reduce greenhouse gas emissions. Although most countries signed the Protocol, the United States and Australia were notable exceptions.

The Kyoto Protocol has generated critiques just as harsh as those generated by *The Limits to Growth*. Critiques from the political left have alleged that the Protocol did not go far enough to fix the global warming threat by capping carbon dioxide emissions at a safe level; critiques from the political right, conversely, have asserted that global warming is part of a natural cycle and is not the result of anthropogenic emissions of greenhouse gases into the atmosphere.

In retrospect, the wording of "tipping point" assessments such as *The Limits to Growth* and the Kyoto Protocol, regardless of their critics, often prove quite valuable for improving basic social understanding of abstract and complex issues by raising critical environmental issues to height-ened levels of public concern.

UNFCCC's COP15, to be held in Copenhagen, Denmark in December 2009, has closed the book on the Kyoto Protocol. But Kyoto will take its place as a stepping stone in the history of attempts to protect the environment from irreparable harm. Most likely, however, the pivotal role and value of Kyoto will be forgotten as happened with *The Limits to Growth*.

Nonetheless, contemporary knowledge is based on knowledge and information supplied by predecessors. One can legitimately ask to what extent climate change concerns of the past few decades – as manifested in the Kyoto Protocol – were sparked by the awareness raising of *The Limits to Growth* almost four decades ago.

Whatever happens at COP15 (in Copenhagen) or at future COPs can likely be directly linked to the debates about the effectiveness of the Kyoto Protocol and possibly to the debates that followed *The Limits to Growth*. COP15 negotiators will have developed guidelines for action for national governments to follow in attempts to avoid a dangerous interference with the global climate system.

What *The Limits to Growth* uncovered in the early 1970s, the Kyoto Protocol set out to highlight in the late 1990s, and what COP15 in 2009 underscored is the following realization: societies are the root cause of profound climate change and impacts on its ill-fated victims.

Notes

Chapter 1 image: "Cyclone Devastated Mangrove Forest", Gregory Pierce, Sundarbans National Park, Bangladesh, February 2009.
Chapter 2 image: "Mountain and Cloudscape", Gregory Pierce, Rocky Mountain National Park, Colorado, USA, July 2008.
Chapter 3 image: "Bitterroot Forest Fire", John McColgan, Bitterroot National Forest, Montana, USA, August 2000.
Chapter 4 image: "Fisherman Casting Net", Gregory Pierce, along the Bhairab River, outside Khulna, Bangladesh, February 2009.
Chapter 5 image: "Bamboo Forest", Gregory Pierce, Lowacherra National Park, Srimangal, Bangladesh, March 2009.

Selected Reading

Boston, Jonathon, ed. *Towards a New Global Climate Treaty.* Wellington: Institute of Policy Studies, 2007.

Burroughs, William James, ed., World Meteorological Organization. *Climate: Into the 21st Century.* Cambridge: Cambridge University Press, 2003.

Glantz, Michael H., ed. *Drought follows the plow.* Cambridge: Cambridge University Press, 1994.

———, ed. *Creeping Environmental Problems and Sustainable Development in the Aral Sea Basin.* Cambridge: Cambridge University Press, 1999.

———. *Climate Affairs: A Primer.* Washington, DC: Island Press, 2003.

———, ed. *Heads up! Early Warning Systems for Climate-, Water- and Weather-Related Hazards.* Tokyo: United Nations University Press, 2009.

Intergovernmental Panel on Climate Change (IPCC). *The Regional Impacts of Climate Change: An Assessment of Vulnerability.* Cambridge: Cambridge University Press, 1998.

Lamb, Hubert H. *Climate, History and the Modern World.* 2nd ed. London: Routledge, 1995.

Mann, Michael E. and Lee R. Kump. *Dire Predictions: Understanding Global Warming.* New York: Dorling Kindersley Limited, 2008.

National Research Council, Committee on Abrupt Climate Change, *Abrupt Climate Change: Inevitable Surprises.* Washington, DC: National Academy Press, 2002.

Shiva, Vandana. *Water Wars: Privatization, Pollution, and Profit.* Cambridge, MA: South End Press, 2002.